Patterns of Emergent Literacy

For Trudie, Sam, Talia, and Harry

Patterns of Emergent Literacy

Processes of Development and Transition

Stuart McNaughton

Melbourne
OXFORD UNIVERSITY PRESS
Oxford   Auckland   New York

OXFORD UNIVERSITY PRESS
Oxford   New York
Athens   Auckland   Bangkok   Bombay
Calcutta   Cape Town   Dar es Salaam   Delhi
Florence   Hong Kong   Istanbul   Karachi
Kuala Lumpur   Madras   Madrid   Melbourne
Mexico City   Nairobi   Paris   Singapore
Taipei   Tokyo   Toronto
and associated companies in
Berlin   Ibadan
OXFORD is a trade mark of Oxford University Press

© Stuart McNaughton, 1995
First published 1995

This book is copyright. Apart from any fair dealing
for the purposes of private study, research, criticism,
or review as permitted under the Copyright Act, no
part may be reproduced, stored in a retrieval system,
or transmitted, in any form or by any means,
electronic, mechanical, photocopying, recording, or
otherwise without prior written permission.
Enquiries to be made to Oxford University Press.

Copying for educational purposes
Where copies of part or the whole of the book are
made under Part VB of the Copyright Act, the law
requires that prescribed procedures be followed. For
information, contact the Copyright Agency Limited.

ISBN   0 19 558324 8

Edited by Maria Jungowska
Indexed by Russell Brooks
Typeset by Scope
Printed in Malaysia by SRM Production Services Sdn Bhd
Published by Oxford University Press,
540 Great South Road, Greenlane
Auckland 5, New Zealand

Mihi

E ngā iwi o te ao,
Tēnā koutou katoa.

Tēnā koutou i ngā tini aituā
Kua mene atu ki te pō.
Nā reira,
Haere, e ngā mate, haere.
Haere i runga i tērā huarahi
Kua whakaritea
E ngā mano, e ngā tini.
Ka āpiti hono, tātai hono,
Koutou ki a koutou
Haere atu rā, haere

Kei ngā mana, kei ngā reo,
Kei ngā iwi, kei ngā whenua,
Tena koutou katoa.
He mihi tēnei ki a koutou,
Ahakoa ko wai,
E mau ana, e hāpai ana
I ngā tikanga hōhonu,
E pupuri ana i te mauri motuhake,
E mahi ana i runga
I tā te rangatira mahi
A koro mā, a kui mā
E tiaki ana i ngā taonga
Kua mahue mai i runga i ngā tipuna.
Tēnā koutou, tēnā koutou, tēnā koutou katoa.

Na Pita Sharples

Acknowledgements

I am indebted to the families with whom I have worked over the last several years. Their willingness to let others see them 'at work' represents a considerable commitment in time, resources, and trust, for which I am very grateful.

I have been nurtured and guided by a number of colleagues who have provided much needed intellectual stimulation and challenge. Parts of the book have been discussed with Courtney Cazden, Marie Clay, Richard Cowan, Ted Glynn, Catherine Snow, and Jaan Valsiner. Timely discussions occurred with research groups at the University of North Carolina at Chapel Hill (Psychology Department), Harvard University (the Graduate School of Education) and at the London Institute of Education (Department of Education and Special Needs) while I was on leave from the University of Auckland in 1993.

Intellectual and emotional support also came from friends in the Research Unit for Maori Education at the University of Auckland, particularly Margie Hohepa, Kuni Jenkins, Judith Simon, Graham Smith, Linda Smith and Pita Sharples. Kia ora te whanau, ka nui te aroha ki a koutou.

This book has depended upon collaborative research with students and with colleagues. I have appreciated and learned from our joint endeavours. The research contributions of Viliami Afeaki, Teresa Ngau Chun, Kathryn Glasswell, Megan Goodridge, Wayne Johnston, Tania Ka'ai, Margaret Kempton, Gwenneth Phillips, Pera Royal-Tangaere, Eileen Taogaga, Lavinia Turoa, and Ema Wolfgramm have been particularly significant. A number of educators have been associated directly and indirectly with the various research reports. Their assistance and professional support has been essential. Among the many, I thank especially Libby Limbrick, John McCaffery, and Sue McLaughlin. Colleagues with whom I have worked in the Education Department at the University of Auckland have influenced the writing of the book. They include Kay Irwin, Dennis Moore, Judy Parr, and Viviane Robinson.

The University of Auckland's generous research and study-leave provisions have enabled me to complete this book. I am thankful to the New Zealand Ministry of Education for funding support for several of the studies reported here. Their continued support is gratefully acknowledged. Similarly, the Ministry advisers on the various projects made timely contributions.

A number of colleagues reviewed parts of the manuscript and offered advice. I thank particularly Richard Cowan, Ted Glynn, Libby Limbrick, and Jaan Valsiner.

I have enjoyed working with Linda Cassells at Oxford University Press. She has provided exceptional advice and her suggestions have made a much better production than would have otherwise eventuated. Maria Jungowska's editing has been excellent. I have learned from, and appreciated her expertise.

Contents

Foreword		ix
Introduction		xi
Part One	**The Wide Lens**	1
Chapter One	Building a Model of Early Development	2
Chapter Two	Resourceful Families	17
Part Two	**Actions and Interactions**	37
Chapter Three	What's in a Name?	38
Chapter Four	A Question of Development	58
Part Three	**Early Activity Systems**	81
Chapter Five	Early Activity Systems: Reading	82
Chapter Six	The Special Case of Reading Storybooks	103
Chapter Seven	Early Activity Systems: Writing	124
Part Four	**Relationships and Transitions**	145
Chapter Eight	What Develops?	146
Chapter Nine	Settings: Home, Early Childhood, and School	161
Chapter Ten	Resourcing Families and Educators	179
Glossary		199
References		202
General Index		211
Author Index		213

Foreword

Patterns of Emergent Literacy has a refreshing approach to the study of parents' interactions with their children that will appeal to readers who are interested in how children develop in early childhood in many parts of the world. It describes some ways in which families introduce children to written language — the stories, the forms, the attitudes — but most important in the author's argument is the discussion of how those families have fostered expertise in their children. One way to explain 'expertise' is to consider that children have been taught 'how to learn' according to the values and concepts of their particular culture.

The book arises from research studies which described how parents from different cultural and language groups in New Zealand interact with their preschool children. It describes different ways in which parents give help to children but these interactions are placed in the broad context of many social and environmental influences on children's learning. The account escapes from the chauvinism of white, English-speaking, middle-class explanations and creates a theory which can explore equally well how and what children learn from both majority and minority cultures in their homes.

Around the world today there is a wealth of understanding of cultural difference: many sources of writing worldwide contribute to the dialogues of cultural movements among different minority and majority groups. However, the arguments for diverse programmes adjusted to small minorities often remain unheard as the relentless tide of research and theory on the majority culture is easy of access and usually influences the guidelines for practice which education ministries hand out to teachers. The 'weight of the evidence' is, in fact, misleading. What we know of cultural differences in minority societies needs to be articulated clearly so that the information has the same validity as the majority culture discussion. This book is important because it helps us to escape from majority-culture bias and provides a common framework for understanding how different processes can result in common outcomes, such as becoming literate.

The author's framework, or discipline, is developmental psychology and the book is designed for educators concerned with literacy in families, early-childhood settings, and the first years of school. It poses a more general question also. If children are actively constructing what they learn, and are the lead actors in anything they learn, then how do we articulate the links between that learning and influence of cultural variables? The issues of today's societies make it impossible to ignore the context in which this constructive learner lives. The author's purpose is to provide 'a coherent theoretical account which examines the nature of, and diversity in, the development of expertise in early reading and writing'. The resulting theory should explain learning for majority-culture children as well as for minority-culture groups, and provide a theoretical tool for exploring different questions.

McNaughton's co-constructivist view of child development can explain literacy development on several levels: the social and cultural functions of family practices of literacy; the presence within practices of literacy activities that reflect social and cultural meanings; the function of activities as systems of learning and development; the development within these systems of specific forms of expertise; and developmental processes at work in the relationships between different socialisation/education settings.

I have been conscious for many years of my endeavours to describe the child's construction of his/her own control over literacy, while this author was saying, 'but that does not take into account the co-construction of literacy in interaction socially with others in the child's home and community contexts'. I know I had a narrowed purpose suited to a different goal and so this framework can serve as a theoretical tool for exploring many aspects of cultural diversity and their contribution to children's learning. We can escape from the straitjacket of socioeconomic classifications of parents in our research and explore the learning practices, activities, systems, and processes occurring in different families.

When we ask questions about school entry, and if we want to use what children bring with them as a starting platform for their learning in school, we have to deal with complex individual, and a wide range of social and cultural differences. The author has constructed an explanatory framework of theory for dealing with such complex issues. Using this theory we can ask questions about how particular children learn in multicultural societies. Given a framework of theory that helps to identify important interactions between children and learning opportunities in their environment, the challenge then is to understand what cultural diversity means for day-to-day interactions in preschools and schools. This book presents good examples of what occurs, and creates a springboard for rethinking good practice in the first years of school. Both experienced classroom teachers and those who train new teachers could draw inspiration from this model for designing better learning opportunities in early childhood.

Before readers take notice of the claims that any authors make from their own interpretation of their own research results they should act as critics and enquire, 'Did they ask important questions?' In this book the questions are at the cutting edge of theory, and deal with the most critical and poorly managed aspect of education in all countries in the world — transition into school, including the first year of schooling. If we get this right, we place preschool experience in the credit ledger, weighing in to support subsequent learning; alternatively we can put most of the child's preschool learning in the debit ledger and proceed to attempt to undo what the child has learned about the world.

One of the important question posed relates to my theory of the construction of the child's inner control for early literacy learning. I have to decide whether McNaughton's view of systems alters the value of that concept in my theory or whether it delineates a limitation in my thinking that can be overcome. So for me, this is a very thought-provoking book.

Professor Marie Clay
Educational Consultant
1995

Introduction

This is a book about how young children learn to do things with written language, from birth up to the beginning of their schooling. It examines their emerging skills with written language in the early years of their lives. It looks at their learning and development in homes and communities, in early-childhood settings, over the transition to school, and in the first few years at school. This book seeks to understand the processes of this learning; to understand what children and their families do with written language; what living in different communities means for literacy; and the part that families, communities, and educational settings play in development.

For as long as I have been researching I have been concerned with three issues. The first is how children learn. The second is how families, early-childhood settings and classrooms contribute to children's learning and development. And the third issue is how children develop in their cultural and social lives before and through schooling.

Partly these concerns came from my early training as a primary school teacher. During that training I kept being confronted by classroom situations with which I didn't know how to deal. Of particular concern to me were the many expressions and levels of language skills presented by children from different cultural groups in New Zealand as they began at school. How to acknowledge their many talents and yet teach them the various forms of expertise valued by the curriculum?

As I have worked with teachers I have learned more about the ways of effective classrooms from them. I have worked also with parents and families, other researchers and with children. I have learned from these experiences too. What I want to discuss in this book is what I have come to understand about how young children learn and the complex relationships between learning and the contexts within which children learn. I have become convinced that in order to create better classrooms for all children we need to understand better what children and their families bring to classrooms.

Writing this book I have had in mind three audiences. I have written it for families who are, as I have learned again, keenly interested in their children's development but often do not have access to professional and academic knowledge. I hope the book increases that access, either directly, or through channels provided by educators who read the book. Educators constitute the second audience. The book has been written to inform and to contribute to more effective educational practices in early-childhood education, in the early years of schooling, and in family education. The research community working in areas of emergent literacy, family socialisation, and early development is a third audience. An aim of the book is to develop a model of emergent literacy that enables us to think and research more productively in these areas.

I have provided guides for these audiences in the form of comments at the end of chapters which highlight ideas relevant to each audience. Another guide is the glossary at the end of the book. Many of the technical and theoretical terms are explained in the text, but some of the more abstract or technical of these have been explained further in the glossary.

Part One

The Wide Lens

Chapter One

Building a model of early development

> **Focus**
>
> **An overview of the ideas in the book**
> - A model for understanding the socialisation of emergent literacy is outlined.
> - The model is summarised in seven propositions about learning and development.
> - The model comes from a theory of co-construction.

A Socialisation Model of Emergent Literacy

The descriptions and explanations offered in this book are in the form of a socialisation model of development. That is, a model is used that focuses on the processes by which children come to be expert members of their families, individuals having special knowledge and ways of behaving. This model incorporates particular psychological explanations of learning and development. These are based on an evolving theory that has been called co-constructionist theory (Valsiner 1988; 1994a).

> Development occurs as children act to make sense of their world, and guidance is provided in everyday activities.

This theory, which is elaborated on throughout the book, argues that development should be seen as occurring through complex and dynamic exchanges between its parts. The parts are, on the one hand, children and their actions to make sense out of their world (their constructions); and, on the other hand, the social and cultural processes in everyday activities, such as what guidance is given, and how it is given. The latter give children's actions form and substance. Development is built from these parts. They are dependent on each other but also part of each other, hence the term co-construction.

Because succeeding chapters use this model based on the psychological processes of co-construction to examine children's development, it is introduced and summarised in this first chapter.

The socialisation model has five components, described in Figure 1.1:

> The socialisation model has five components: family literacy practices, activities, systems for learning and development, expertise, and relationships between settings.

1. The family's **practices** of literacy are the general purposes and uses of written language that enable the family to function. These practices, such as maintaining contact with distant family though exchange of letters, form an enveloping framework of social and cultural meanings shown as an outer (prominent) circle.
2. Three sorts of **activities** in which children and family members engage are encompassed within the family's practices — **ambient** (surrounding), **joint**, and **personal** activities. These activities are selected, arranged, and deployed (implemented) by families.
3. The ambient, joint, and personal activities create opportunities for the formation of **systems** for learning and development to occur, whose parts comprise the activity, the learner, and family members. As with all systems, these parts interact, providing the sites where co-construction takes place.
4. Out of these systems come specific forms of **expertise** which are situated in the activities. Reading books to children, for instance, involves the use of family resources. An activity is created which contributes to family practices with written language, such as using books for recreational purposes. The progressive interactions that take place within the activity create a system which enables the learner to become an expert, which might entail strategies for understanding written stories.

Figure 1.1 A socialisation model of emergent literacy

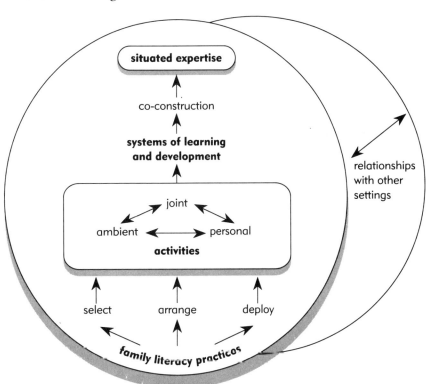

5. Practices, activities, and systems are present in other socialisation settings too. The last part of the model concerns developmental processes at work in the **relationships** between settings. I introduce each of the model's five components before discussing them in detail in the following chapters.

1 Family practices

Families are often described as environments for children's development. But they are unusual environments, unusual because family members are both the environment and (at least partly) the creators of the environment. A better representation of the family for the purposes of exploring its developmental properties uses the general concept of a system (Sameroff 1982).

As in other dynamic systems the family functions to achieve certain purposes. Systems respond to outside and internal events to maintain and develop these goals. One of the central functions of the family is to socialise children into the ways of thinking and acting which are appropriate for the community (or communities) of which the family is a member. This is what counts as expertise.

> One of the central functions of the family is to socialise children into the ways of thinking and acting which are appropriate for the community of which the family is a member.

In developed countries such as New Zealand the socialisation goals that families have for their children include becoming literate. Family members use written language for a range of purposes at home as well as away from home. The purposes may be closely tied to particular occupations or responsibilities. For example, parents who are secretaries of local sports clubs may have the job of writing the minutes of meetings, sometimes doing this at home as well as at the club. Family members may write to relatives living elsewhere. Brothers or sisters may bring home books or other forms of homework from school. Menus at fast-food restaurants may be read as the family places orders. All of these are expressions of the literacy practices of the family and through them children are socialised into literacy.

> The purposes of literacy events and the forms they take identify members of the family in social and cultural terms. What one does with written language both reflects and expresses one's identifying characteristics.

The purposes of these literacy events and the forms that they take identify members of the family in social and cultural terms. What one does with written language both reflects and expresses one's identifying characteristics. For example, it identifies the groups to which you belong and the sorts of work you do. This identification factor is an integral part of the understanding of family literacy. This point is taken up in the next chapter. It is why the socialisation model identifies social and cultural messages within literacy practices. The model is based on a view of culture as shared meanings which underlie ways of acting in particular contexts. In this view, culture resides in everyday practices.

Young children are often on the periphery of these purposeful events. But, as I will show, even on the periphery they are very active spectators. In yet other events children are centrally involved. So there are occasions when children are directly engaged with someone else in an event where the medium is written language. Menus might be read to children. Books might be read to children. An alphabet might be sung and pointed at. Part of rearing children in literate societies involves families socialising children directly into ways with written words.

Additional events are created by children as they play and work with written

BUILDING A MODEL OF EARLY DEVELOPMENT

language by themselves. All of these events, the ones in which children are not directly engaged and the ones in which children are directly engaged either with others or by themselves, contribute to the social and cultural purposes of the family.

> All written-language events contribute to the social and cultural purposes of the family.

The major claims in the socialisation model with which different chapters deal are summarised in seven propositions. Chapter 2 explores the first two of these propositions which are about literacy and family practices:

Proposition One: Families arrange time and provide resources which socialise children into their practices of literacy.

Proposition Two: Family literacy practices reflect and build social and cultural identities.

2 Activities

How a family practises literacy can be seen in particular events of reading and writing, which are described in theoretical terms as activities. Chapters 5, 6, and 7 systematically define these various activities. They are the habitual family events which children experience as observers, as participants, and when they play and experiment by themselves. Although the concept of activity has been subjected to detailed theoretical analysis by writers in the field, the activities are so familiar that Tharp and Gallimore (1988) describe them as 'the social furniture of our family, community, and work lives' (p. 72).

> Reading and writing activities have specific goals and recognisable structures seen in the patterns of behaviour and the means of carrying out the activity.

Activities have specific goals and recognisable structures seen in the patterns of behaviour and the means of carrying out the activity. For example, in some families, writing letters to relatives and friends often occurs. The activity of letter writing has goals, some obvious and others not so obvious, but nevertheless identifiable with careful examination. In our family sometimes we have written letters as a means of occupying time while waiting with preschool

Figure 1.2 Three types of family literacy activities

1. Joint activity: a father and two siblings reading a book with a 3-month-old child.

2. Ambient activity: the child's two siblings writing and reading.

3. Personal activity: the child at 9 months reading a book alone.

children for a consultation with the doctor. Who carries out the activity in a family, and how and when it is accomplished are important features. These are features of the structure, or rules of the activity.

Activities are the mechanisms in the family system which achieve family purposes and express their social and cultural identities. Family members may have thought carefully about these functions and how they meet the needs of the family system and may even be able to describe and explain them clearly. But they might also be so habitual or so deeply ingrained in everyday life that the need to think 'out loud' about purposes may have never occurred. Nevertheless,

> Activities are the mechanisms in the family system which achieve family purposes and express their social and cultural identities.

activities are meaningful and participants have, or develop, motives for participating. I will be stressing how activities have several goals within the family.

Chapters 5, 6, and 7 are devoted to examining different sorts of activities for reading and writing. In each chapter specific activities are identified. The rationale behind these chapters is contained in a further proposition.

Proposition Three: Literacy practices are expressed in specific activities which have identifiable constituents. These include goals, rules for participation, and ways of carrying out the activity.

Proposition Three is based on a more general one which is hardly contentious now. It is that the development of literacy begins a long while before school. It is useful to pause here and note explicitly what this claim means. During the 1980s I wrote a book about learning to read (McNaughton 1987). I started writing it not long after some developmental writers had argued that reading was not a developmental phenomenon. That is, it had been argued that because reading was something that had to be taught (at school) therefore it was not something that developmentalists should properly spend time trying to explain.

It was not difficult to argue against this view. Descriptions of children were showing the presence of substantial amounts of learning before school (Clay 1979; Teale 1984). Much of what has been written in the last decade clearly shows that reading and writing have developmental roots. The term 'emergent literacy' has come to refer to this early learning.

It is appropriate that I use the term in this book. One reason for it being appropriate is an acknowledgement of the pioneering work of developmental psychologist Marie Clay, who was one of my mentors at the University of Auckland. In her early research she introduced the concept of new skills continually emerging during development (Clay 1966).

I use the term emergent literacy to convey several ideas. One is that expertise in reading and writing has a developmental history before formal instruction. That expertise emerges from processes at work in children's everyday experience. Along with other writers, I use the term literacy rather than reading and writing because it 'signals a recognition of the complex relationships among reading, writing, ways of talking, ways of learning, and ways of knowing. Literacy is not just a cognitive achievement on the part of the child; it is also participation in culturally defined structures of knowledge and communication. [It] means achieving membership in a culture' (Snow *et al.* 1991; p. 175).

This definition is wider than some others. For example, Sulzby and Teale (1991) adopt the definition of emergent literacy as the reading and writing behaviours that precede and develop into conventional literacy. The need for a wider definition follows from my concern to understand social and cultural processes. The concept of literacy employed by such writers as Snow and her colleagues stresses the need to understand the diverse uses of symbol systems, rather than those forms of reading and writing that develop into conventional school uses.

The research on emergent literacy has established development before formal instruction. Now it has to try to explain how development has occurred, seemingly in the absence of explicit teaching. As in studies of language development

---

> The development of literacy begins a long while before a child starts school.

> In the term emergent literacy — emergent refers to the developmental roots of literacy before school.

> The term literacy 'signals a recognition of the complex relationships among reading, writing, ways of talking, ways of learning, and ways of knowing'.

this has meant the roles of socialisation processes in development and learning have been hotly debated. Different theoretical approaches to this issue are introduced shortly and discussed further in Chapter 4. In common with many other writers the approach taken in this book is that instruction at home is not like typical school forms of instruction. But unlike some other approaches, this book proposes that direct and active ways of contributing to development certainly take place. They are very obvious in family activities.

3 Systems for learning and development

> Activities which involve children provide a framework for learning, but a learning and development system has to take shape within the family to activate the framework.

The activities which involve children provide a context or a framework for learning. But something has to take place for the framework to be activated to produce learning and the development of expertise in family literacy. That something is called a learning and development system. Such a system takes shape within the family, which has already been described as a system, too. Learning and development systems can be seen as subsystems within the larger one.

Personal and social forces combine to create a learning and development system. One force is provided by the child as an active problem solver. Under appropriate conditions children set out to solve the problem of how to do what others do or expect them to do with written language. The other force is carried in the moment by moment interactions between the child and family members, and others with whom the child interacts regularly.

> Systems develop in activities that directly involve family members.

Two sorts of systems are described. There are systems which develop in activities that directly involve family members. For example, in some families books that contain storylines (narrative books) are read with children. The interactions that take place during this activity and the ways in which these develop over time illustrate this first type of interactive system. Interestingly, there are different ways that storybooks can be read with children and this means that this basic inter-active type of learning and development system can come in different forms (see chapters 5 and 6).

> Experimentation reflects a child's social experiences.

A second complementary system occurs as children play and experiment with reading and writing by themselves. Some children play at reading a storybook to themselves, even pointing to things on the pages. Some play at writing on paper and make intricate lines and shapes. In the process they are practising and instructing themselves about reading storybooks or writing. The systems which develop here are operated by the children themselves. But they owe their genesis and developmental power to the presence of the first type of system. When 2-year-olds scribble by themselves and even when a 1-year-old picks up a book and tries to eat it, their experimentation reflects their social experiences. This is why even these independent sorts of learning and development systems can be said to represent the coming together of two forces.

These ideas about learning and development systems can be summarised in two further propositions:

Proposition Four: Systems for learning and for development take form within activities as a product of the child's actions and the actions of significant others.

Proposition Five: Two basic and complementary types of system occur and each can be expressed in a number of ways.

4 Situated expertise

Development occurs within activities given that a learning and development system takes shape. What develops are forms of expertise which are situated in those activities. That is, children come to be expert in the activities within which they participate. For example, some children develop expertise within the activity of storybook reading. It is now possible to be quite specific about the nature of this expertise. Of those children who participate in storybook reading there are some who become expert at retaining and repeating segments of books as a consequence of the particular ways in which readers and listeners have interacted. For others the interactions provide them with the expertise to focus on the possible meanings in books that are being read to them (see Chapter 6).

> Children come to be expert in the activities within which they participate.

I have carefully chosen the phrase 'situated expertise', which is used in the socialisation model (p. 3) and follows the ideas developed by other writers such as Wozniak and Fischer (1993) and Rogoff and Lave (1984). It means that the form of the expertise reflects the form that the activity has taken. This argument is examined in the descriptions of activities and expertise contained in chapters 5, 6, and 7. One implication of this view is that different sorts of expertise, and therefore different expressions of literacy, develop before school.

Of course, certain properties of written language and psychological processes are common to the different forms that expertise might take. Listening strategies with biblical texts might differ from those associated with some sorts of storybook reading. But commonalities also exist in the processes of listening. Similarly, the discriminations associated with learning an alphabet are different from but have commonalities with the discriminations involved in writing a name.

Depending on the nature and range of these activities, the interaction processes, and the actions of the child, the commonalities may become more or less obvious. These will determine the degree to which multifaceted skills develop that support and reinforce one another.

The general model of socialisation adopted in this book describes development in terms of the notion of expertise (Wood 1988). This way of looking at what develops is used in the book because the concept of expertise means problems in other sorts of descriptions can be avoided. From this viewpoint development is seen as a shift from being a novice to being someone with facility in a particular activity.

> Becoming an expert involves gaining the meaning of the activities, which includes their purposes and values, and how to participate.

This enables me to use a social and cultural frame of reference. Becoming an expert involves gaining the meaning of the activities, which includes the purposes, values, and roles that are part of the activity. In this way, a child who becomes expert at listening to stories being read is learning, among other things, what storybooks stand for, what purposes are served by them, who reads them, how they are read, and so on. Other social and cultural meanings include family roles such as whether or not older siblings, or extended family members read to young ones.

> Socialisation is how the novice develops family-based expertise. What develops is more than aggregates of behaviours, more than an expression of general intellectual development.

Socialisation is central to the development of family-based expertise. This is how the novice gains the expertise of others in the family. The shifts in expertise go from a relative dependence on the supports provided through learning and development systems in the family, to relative independence from those supports. What develops is more than aggregates of behaviour. And it is more than an expression of general intellectual development.

The concept of expertise draws attention to the development and integration of skilful behaviours. For example, a 4-year-old who has often participated in story reading in which the story is the focus of interactions develops some knowledge about the nature of stories. He or she develops some strategies for comprehending stories which are read to them. It is also likely that the well-practised 4-year-old listener has some sophisticated ways for figuring out if they are listening to a story that makes sense. This involves ways of checking and problem solving during the activity.

These meanings of expertise are summarised in Proposition Six and are explored in detail in chapters 5, 6, 7, and 8.

Proposition Six: What children learn to do with written language is become relative experts within particular activities.

5 Relationships between settings

Families are not the only setting where children may encounter practices, activities and systems for learning and development in which expertise in written language develops. More formal educational settings also socialise children, so that situated expertise develops in reading and writing. Other significant settings, including churches and clubs, may also provide opportunities for ambient, joint, and personal activities.

The presence of several settings in which developmental processes may be at work raises important questions. Does the relationship between the practices, activities, and systems in these different settings make a difference to the development of expertise in the settings? How situated is expertise, and can connections be made between situations which influence further development? If such connections are possible, how might systems be better co-ordinated?

> Families are parts of a wider environment incorporating other socialisation settings. The degree to which developmental processes are closely related in different settings makes a difference to the development of expertise.

Two major concepts are added to the socialisation model to answer these questions. The first is the concept that families are parts of a wider environment incorporating other socialisation settings. The second is that processes can be identified which are at work in the relationships between settings. Each of these concepts is partly based on the theoretical contributions that developmentalist Urie Bronfenbrenner has made in his work on the ecology of human development (see Bronfenbrenner 1979; 1986). The socialisation model needs to include these concepts and I discuss relationships between literacy in the family and the wider environment in Chapter 9.

It appears that the degree to which developmental processes are closely related in different settings does make a difference to the development of expertise. And this is not just an additive effect. That is, children's development can be markedly enhanced in each of the socialisation settings if the settings are well co-ordinated. For example, children's early progress in reading at school can be accelerated by family members hearing children reading school texts at home. Apart from how often this occurs, it matters what sorts of texts children read and how family members interact with the reader when hearing the reading. The detailed explanations for this example, which illustrate the significance of relationships between practices, activities and systems of learning and development between settings, are described in chapters 9 and 10.

The co-ordination between settings can be carried in the products of activities. As I have noted already, participants in activities develop a range of goals and ideas in these activities. My 4-year-old daughter painted the picture shown below at her kindergarten. Her bringing it home contributed to the relationships between our activities at home and the kindergarten.

Sample 1.1 Talia's painting and her text. The product resulted from joint activity involving Talia and her kindergarten teacher. By bringing the painting home, Talia contributed to the developmental processes connecting the home and the kindergarten

Talia's production with the text was a collaboration with a kindergarten teacher. Her contribution was the painting and the oral commentary, which was expressed like a written text. She and her teacher collaborated in producing a form of written language and it included Talia naming the production (Talia). The product that came home informed and fed into our socialisation practices at home. Among other things it told us that Talia may be interested in and capable of writing her name collaboratively at home as well as in the early-childhood setting. It told us about her development in creating written texts. But it also told us something about the messages she wanted to write. Actually, in this case the production of the text followed the painting. She had started by painting my face and when the paint for the eyes 'ran' she created an apposite text. I had not been aware of feeling sad at being at work.

Close literacy relationships are based on connections between settings. These include formal links, such as planned parent–teacher meetings, and informal links, such as the sharing of writing activities between kindergarten and home. Hearing children read books at home illustrates some of these informal links, particularly the degree of shared knowledge. What family members know about written language activities in the classroom and what teachers know about what happens with written language at home are among the determining factors for close and effective relationships. But there are also boundary conditions. Relationships are enhanced under circumstances where the family's community identifies with the school. This identification may take

> Close literacy relationships are based on connections between settings. What families and teachers know about written language activities in their respective settings are among the determinants of effective relationships.

place through having some control or functional participation in school decisions and activities.

These ideas about relationships are summarised in the last proposition and analysed in detail in the last two chapters.

Proposition Seven: Development is enhanced by the degree to which settings are well co-ordinated in terms of practices, activities, and systems of learning and development. This in turn depends on a number of boundary conditions, including how participants in settings relate to one another.

> Development is enhanced by the degree to which settings are co-ordinated.

Opening the Lens: A Theoretical Note

Models such as the one described here draw on basic psychological concepts. It is important to identify these theoretical concepts. The descriptions and explanations offered in this book should become clearer (and more able to be queried and checked) if the lens of enquiry is opened to include those broader concepts.

One of my purposes in writing this book on early literacy is to contribute to ongoing theoretical debates in the contributing disciplines of developmental and educational psychology. One theoretical objective is concerned with how to understand the role of culture in models of development. A related objective is to contribute to elaborating on the co-construction theory of development. In the last part of this introductory chapter these objectives are outlined.

The legacy of Piaget and constructionism

David Wood (1988) describes how several theoretical traditions have contributed the great explanatory themes for how children think and learn. These themes are present in studies of emergent literacy, too.

One of these traditions is the extraordinary contribution to constructivist theory of psychologist Jean Piaget. Piaget created a theory of cognitive development which, among other things, described children as going through qualitatively different stages of thinking about their world (Piaget 1970). As children grow older they develop new ways of constructing knowledge and interacting with that world. Constructivist theory is still strongly advocated in developmental theory generally (Carey & Gelman 1991), and in the field of emergent literacy (Goodman 1990).

But there are strong criticisms of Piaget's formulation. These criticisms have shown that some of the central assumptions and explanations of his and others' findings were wrong. For example, young children (preschoolers) are not as illogical and, as it were, single-minded as Piaget's stage theory would suggest. Moreover, children can show qualitatively different forms of reasoning under different conditions, essentially being at different stages at the same time (Morss 1991; Wood 1988).

> Children can show qualitatively different forms of reasoning under different conditions, essentially being at different stages at the same time.

I do not survey the detailed criticisms of Piagetian theory here, but rather acknowledge the legacies of Piaget's constructionism that are present in the socialisation model outlined earlier. There are four underlying concepts that are particularly significant to the model. The first is the notion of the child as an active strategic constructor of knowledge.

Although I do not share Piaget's view that children's actions are the primary source of development, the idea that children are active in learning is nevertheless fundamental to the model. Representing children's development as gaining expertise entails ideas additional to those of Piaget. But the basic assumption of the active learner is present when the child's actions are described as strategic, (that is, adaptable and flexible to meet different demands), regulated (that is, controlled to maintain effective action), and knowledgeable (that is, drawing on a particular base of knowledge about the expertise).

Another legacy of Piaget's work is the idea that learning takes place as children confront and solve problems. They act on problems in that they are motivated to make sense of their worlds, to reduce ambiguity and uncertainty, and to become more expert in their immediate environments. An implication of Piaget's view provides an important challenge for understanding children's actions on problems. The nature of the problem, in particular how it is seen and the ways to a solution, has to be understood from the child's perspective. What an adult expert might describe as the problem to be solved in learning how to write one's name, for example, might not be the child's.

These concepts, that children act in problem spaces and they understand (represent) them in their own childish ways, carry a further implication. It concerns the roles and meanings of children's mistakes. When a child writes MUM in big letters covering a whole page, for example, it may say something about what that child assumes about writing. It might indicate that the child thinks that big and important people need letters that are big and encompassing (see Czerniewska 1992). When a child looks at a picture and makes up a story based on it that sounds like book language, the 'error' (there is no actual text written) tells an observer important things about the child's developing knowledge.

Each of these concepts about children's development underlie the descriptions and explanations of expertise in the following chapters. An anecdotal observation serves to illustrate them here. When my youngest son Harry was 3 years old he made the following observation about the nature of reading. He and his grandfather (who was a regular caregiver) were preparing to have lunch. His grandfather had made him a sandwich out of a yeast extract called Marmite. This brand name is displayed prominently on the jar. There is a very similar vegetable extract that his grandfather prefers and often uses as a generic name for both. The alternative is called Vegemite. Having made the sandwich Harry's grandfather said, 'Here's a Vegemite sandwich.' Harry queried saying, 'No it's not. That's an /M/ for Marmite.' He then paused and reflected, 'That's called reading.'

An analysis of this exchange draws on Piaget's four concepts. This exchange shows an active and strategic problem solver at work. Harry's query is easily understood as a resolution of conflict between different sources of information (there was no /V/ on the jar, yet his grandfather's label used a /V/). The query and his reflection begin to show what he knows about the nature of 'reading'. and his current strategies (using beginning letters to stand for whole words). With other information they also show what knowledge he has yet to build

> A basic assumption about children's nature is that they are active learners. This lies behind the description of children's actions as strategic, regulated, and knowledgeable.
> Learning takes place as children confront and solve problems.

> How a problem is seen and the ways to a solution have to be understood from the child's perspective.

(reading is more than identifying beginning letters). These ways of looking at the child are legacies of the psychological tradition which owes so much to Piaget's contribution.

> Piaget's theory under-represents the role of cultural and social forces in learning and development.

The greatest inadequacy in Piaget's theory concerns the under-representation of the role of cultural and social forces in learning and development. Simply stated, his view was that social agents provided the raw experiences which fuelled the child's actions, which in turn constructed knowledge. The provision occurred indirectly by making available physical resources or directly in language exchanges. In this view culture is the storehouse of fuel. According to Piaget, different cultures provide more or less of the fuel for children's development. This view is clearly evident in writings on emergent literacy that have adopted a Piagetian perspective (e.g. Goodman 1990).

Co-construction of development

The alternative view adopted in this book is that the bases of development are social and cultural as well as personal. The activities which involve children, from the feeding and playing activities with newborns to reading books with 4-year-olds are the frameworks for learning. The interactions which are part of the activity help structure ways of doing and, subsequently, ways of thinking. In this view, socialisation is active. Learning and development which derive from that learning are co-constructed.

> An alternative view is that interactions which are part of an activity help to structure ways of doing and, subsequently, ways of thinking. In this view, socialisation is active, and learning and development are co-constructed.

The exchange between Harry and his grandfather is fundamentally social and cultural. The brief flash of expertise only makes sense when seen in those terms. This is illustrated at many levels. One concerns the social organisation of the setting. The negotiation of what the writing said, in which the 3-year-old queried his elder, shows what had been learned about the purposes and values of reading. It appears that this included using written language to check whether very significant others are 'right'. Indeed, the dialogue pattern shows important values and beliefs about negotiation and authority.

Yet another level at which this little event needs to be understood with reference to social and cultural processes concerns where the insightful comment *'an M for Marmite'* came from. A co-construction theory provides a framework for explaining the sources of this expertise. The essential ingredients of the framework have been introduced here and the rest of the book elaborates on them.

> The co-constructivist view is a contemporary development which has elaborated on earlier concepts.

The broad concepts which underpin the co-construction theory include those legacies of Piaget's constructivism introduced earlier. In addition, the theory introduces a number of further concepts, three of which are highlighted below. These further concepts can be traced back to the earlier psychological theorising by Vygotsky (1978) and others (see Valsiner 1994a for an historical account). But the co-constructivist view is a contemporary development which has elaborated on these earlier concepts still further (see Rogoff 1990; Valsiner, in press; Valsiner & Van der Veer 1993; Wertsch 1991).

The first of these concepts is that the child's construction of knowledge and, more broadly, the child's expertise in action is created first in and through social interactions. The properties of these encounters construct a basis for the learner's subsequent personal psychological functioning. In this way interactions lead development; or to put it another way, learning which is socially mediated extends independent or personal development.

> Learning which is socially mediated extends independent, or personal, development.

A further critical concept is that cultural and social meanings are expressed and constructed within interactions. Both the more expert others with whom one directly and indirectly interacts, and the learner bring personal meanings to, and develop them from, interactions.

These interactions are not the only source of messages about identity and expertise. Messages about expertise and identity within the family, community, and, more generally, within one's culture are carried in many sources. Different people, different activities, different media, and different settings are all sources. But personalised meanings about ways of being are actively constructed from these sources.

A third concept is that the process of construction is channelled by one's own development and the significant others in one's life. That is, the environment within which development takes place is organised or structured. Particular activities take place which have been selected and deployed. The activities which the learner selects and engages in also contribute to the channelling of their own development.

In this way, what a child is able to do with written language is built within the activities — those jointly engaged in with family members and those which are observed or played with independently. The strategic planful and knowledgeable characteristics of the child's expertise are co-constructed within these activities. They develop through the learner appropriating features of the social experience as his or her own ways of functioning. Barbara Rogoff (1993) uses the term appropriating to mean taking and transforming the interactional processes as one's own. But it is the activity and the patterns of interacting within the activity through which literacy expertise is formed and channelled. In this view the environment (although this is not a useful way to picture the social and cultural framework) is very active. Families more than provide the fuel for development. Together with their children they are both the fuel and the vehicle.

> The child's personal action in reading and writing is a fundamental route to greater expertise.

This view does not exclude or undervalue the developmental significance of play and personal exploration. I argue in chapters 2 and 3 that the child's personal action in reading and writing is a fundamental route to greater expertise. (This is a long-standing principle in the psychology of learning and development that is held almost universally, see Valsiner 1994a). Nevertheless, the argument also will be made that what children play and experiment with in personal activities is closely linked with activities they have seen or in which they have engaged.

One last implication of this view needs to be noted before these concepts are applied to literacy in the next chapter. It concerns the very nature of development. In Piaget's view children develop through a fixed and *unitary sequence* of development. The co-constructivist theory does not assume this. It is possible that expertise with written language can take various forms and that there are different routes to the same expertise (see Scribner & Cole 1981; Gee 1990). Given that expertise can take different forms, different developmental sequences can exist for different contexts of use.

The rest of this book provides the details of socialisation and the processes of co-constructing expertise with written language. The analysis goes into families' lives and beyond, to activities in different settings at church, at preschool, and at school. Drawing the descriptions and explanations together into a coherent whole is the task I have set in this book. I have attempted to provide a general account to serve the wider purpose of trying to make a difference to the quality of children's socialisations into written language.

In the following chapters I use notes signalled in the text to achieve two purposes. One is to avoid cluttering up the discussions in the text with the details of empirical evidence. I add or elaborate on such details in notes at the end of each chapter. The second purpose is to extend theoretical points in the text. Given that I wish this book to contribute to the development of theory, the notes provided at the end are a vehicle for theoretical commentary.

Implications

For families, educators, and researchers

Also at the end of each chapter are notes summarising specific implications the chapter holds for families, educators, and researchers. Because this is an overview chapter I simply outline the general implications of the socialisation model.

Families: Families have a right to educational knowledge and guidance which they can use to inform their socialisation practices.

Educators: Literacy programmes (both relatively formal and informal) need to be based on knowledge of the diverse expressions of socialisation and expertise in families, and the ways in which family literacy and educational practices can influence each other.

Researchers: The socialisation model and its propositions are a guide for the descriptions and explanation of emergent literacy, and for analyses of educational programmes. There are implications for both theory and research practice in each of the propositions.

Further Reading

Further discussion of the psychological explanations which underpin the socialisation model can be found in several sources.

Rogoff, B. (1990). *Apprenticeship in Thinking: Cognitive Development in Social Context.* Oxford University Press, Oxford.

Tharp, R. G. & Gallimore, R. (1988). *Rousing Minds to Life: Teaching, Learning and Schooling in Social Context.* Cambridge University Press, Cambridge.

Valsiner, J. (1987). *Culture and the Development of Children's Action.* Wiley, Chichester.

Wertsch, J. V. (1991). *Voices of the Mind: A Sociocultural Approach to Mediated Action.* Harvard University Press, Cambridge, MA.

Wood, D. (1988). *How Children Think and Learn.* Basil Blackwell, London.

Chapter Two

Resourceful families

> **Focus**
>
> **The role of the family system in emergent literacy**
> - Families arrange time and provide resources which socialise children into their practices of literacy.
> - Family practices reflect and build social and cultural identities.
> - Literacy practices are expressed in specific activities which have identifiable constituents. These include goals, rules for participation, and ways of carrying out the activities.

Families, Child Rearing, and Literacy

In this chapter I explore emergent literacy at the first level described in the socialisation model on page 3. At the heart of the model is a view of the family as a system which has a variety of functions.

One of these functions is child rearing. That is, the family is a vehicle of socialisation through which young members become expert members of the social and cultural groups to which the family belongs.

The family is a vehicle of socialisation through which young members become expert members of the social and cultural groups to which the family belongs.

Members of the family practise literacy. They use written language for purposes that have to do with their roles within and outside of the family. By means of these uses families socialise their children into ways of using written language that are appropriate for the family.

The socialisation model employs the active verbs of **selecting**, **arranging**, and **deploying** to describe the role of the family system in socialising children. Family practices actively channel children's development through the creation of sets of experiences and opportunities. Valsiner (1987) describes these as Zones of Freedom of Movement (ZFM), and within them are Zones of Promoted Actions

(ZPA). The ZFM refers to those aspects of the environment which are organised to enable access by children. The ZPA refers to the particular objects and areas and ways of acting that are promoted or highlighted within the ZFM. By selecting, arranging, and deploying particular experiences and opportunities, families provide their children with certain boundaries within which they can move, both physically and psychologically speaking. Within these zones some particular experiences and opportunities are placed in the foreground.

Actions can be promoted in a direct fashion. This occurs when family members arrange activities for children in which the *children's* interests and needs are the focus. Resources are selected and deployed which result in family members doing things together with their children.

My use of the term resources conveys the sense that child rearing is resource*ful*. Among other things it takes time and effort to read to children or to sing an alphabet chart. Besides time and effort there are other psychological resources that families with different social and cultural identities may bring to the tasks of child rearing. For example, in many studies white middle-class families are described as owning large numbers of books for their children (e.g. Elley 1992).

French sociologist Bourdieu (1973) sees both this physical resourcing and the more complex forms of knowledge and skills constructed in child rearing as providing children with 'cultural capital'. The significance of this way of looking at resourcing is that schools may only recognise some forms of 'capital'. Another way to put this, although I find the metaphor demeaning to both families and schools, is that only some forms of capital may be bankable at schools.

One further way in which families may have resources can be seen in the day to day life of the family as a system. Fox (1990) describes those conditions of life that enable the family system to be 'robust', including such things as networks of support and the time family members have to be available to children. A family that is robust in these terms has the resources to carry out child rearing tasks effectively. A family that is relatively isolated and unable to function well in the midst of social and economic hardship may not find it easy to carry out their child-rearing tasks (see also Gallimore & Goldenberg 1993; Snow *et al.* 1991).

Resources for child rearing are more obvious perhaps when children are in the foreground of the activity, directly participating. There is a second form of socialisation which is less direct and sometimes less obvious. The second route is grounded in what members of the family do every day or what the family environment provides in the way of opportunities for personal exploration. In the former case the child is a peripheral but active observer. In the latter case the child has opportunities to practise and play with materials using their emerging expertise.

Although indirect, this second route is no less active or resourceful than the first. Families select, arrange, and deploy resources for reading and writing to be observed; or for a child to try to read and write by themselves. Again, families can bring different resources which constrain the opportunities available to observe, play, and practise with written language.

This chapter sets out two claims in the model. I examine the first proposition that families engineer, both incidentally and deliberately, occasions for young

> Occasions to learn about written language reflect and construct social and cultural messages for children. From these they develop ideas and values about literacy practices and activities, and their personal and cultural identity.

children to learn about written language (see Chapter 1, p. 5). I also examine the second proposition that these occasions reflect and construct important social and cultural messages (see Chapter 1, p. 5). From these messages children develop ideas and values about literacy practices and activities and, more generally, about their personal and cultural identity. The messages include who carries out an activity, what one does in the activity, when it should take place, why one does it, and even how it should be learned. These ideas and values are interwoven with others creating a network of meanings during socialisation.[1]

A study of socialisations of literacy (SOL)

Rather than review the extensive research which examines the two propositions, the issues can be appreciated in a more immediate way through one illustrative study. The detailed argument is made using a study of 17 families in New Zealand. As it focused on the socialisation of literacy I refer to it as the SOL study. At strategic points I sketch the obvious connections with the more general research literature available from other countries. This is in order to show which parts of the example are generalisable and which parts may have specific significance. To do justice to the example, information about how New Zealand families in general select, arrange, and deploy literacy experiences will need to be reviewed. Again, connections are made with descriptions from studies carried out in other countries.

> The deliberate use of descriptions of emergent literacy in the New Zealand context is consistent with a major theme of this book: literacy develops in particular contexts.

The deliberate use of descriptions of emergent literacy in the New Zealand context serves a wider purpose. Oscar Wilde said that consistency was the last refuge of the unimaginative. I want to be unimaginative here. It is consistent with a major theme of this book that I should attempt to describe the development of literacy in a particular context.

In 1987 our research group at the University of Auckland began a longitudinal (two-year) study of 17 children. It was similar in design and concept to other studies that had been reported (e.g. Wells 1985) or were in process (e.g. Snow 1991). We examined the children's development from the time they were 4 to 6 years old. Approximately every six months we observed closely in their homes and later at school. We asked questions of their families, particularly their parents, and we described their burgeoning expertise with written language.

This project was distinctive in that the children were selected as potential high-progress readers at several schools. This prediction was made on the performance of siblings already at school. In this sense it is an extreme example which may enable us to see developmental mechanisms more obviously. Details of the methodology and research techniques are described elsewhere (see McNaughton & Ka'ai 1990), but two characteristics of the families are important to highlight. The families came from three cultural groups in Auckland. There were four Maori, six Pakeha, and seven Samoan families.[2] Secondly, the earners in the family were in non-professional occupations. Technically the median socioeconomic level for the group was level four of Johnston's (1983) six-level index; typical occupations were machinist, cashier, carpenter, storeperson.[3]

Practices of literacy: the child's view

Families, or more accurately different members of a family, engage in literacy practices. The technical meaning of this term is provided by the Laboratory of Comparative Human Cognition (1983). They describe a practice as involving:

> ... activities for which the culture has normative expectations of the form, manner and order of conducting repeated or customary actions requiring specified skills and knowledge ... Cultural practices have to be learned as systems of activity. (p. 333).

> A family's practices provide a zone within which particular activities are promoted. Three sorts of activities can be distinguished: joint, personal, and ambient.
>
> Ambient activities surround the child and provide a context for jointly constructed and personal activities.

From the point of view of the developing child a family's practices provide a zone (the ZFM) within which particular activities are promoted (ZPA). Three sorts of activities can be distinguished. The first are literacy activities in which the child is engaged directly. These are **jointly constructed** activities. Secondly, there are the times of **personal** or independent engagement with literacy activities. Finally, there are those that make up the **ambient** literacy environment of the child. There is a close association between each of these which will be explored further in ensuing chapters. The ambient activities surround the child and provide a context for the other activities so, I describe these first.

Ambient activities

Older, more expert members of the family do things with written language which preschool children can observe. In the course of their daily lives they expose children to uses of written language. The child's role in relationship to written language activity is peripheral in these experiences. But from this surrounding the child can construct knowledge about ways of doing things with written language.

Figure 2.1 Literacy activities: ambient activities surround the child and provide a context for joint and personal activities

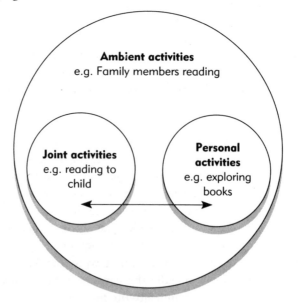

Children engage directly in activities with others (joint) or by themselves (personal) surrounded by activities (ambient) in Family Practices

The knowledge which children construct is incorporated and elaborated on in the other activities, those carried out jointly with others, and those done independently. In turn, the expertise developing in these activities provides a knowledge base for being able to see in more detail that others are reading and writing, and for subsequent constructive imitation of what they are doing. How powerful these ambient activities are as models for reading and writing depends, therefore, on what children are learning in other activities. And what children learn in other activities is affected by the presence of ambient activities which serve as salient models.

> The role of ambient activities as models for reading and writing depends on what children are learning in other activities.

The socialisation model describes these experiences as expressing and constructing ideas and values about literacy in the first instance but more generally about one's identity. This is a claim that such an area of experience is culturally and socially structured. What evidence is there for this assertion? To what extent did the families in our study select, arrange, and deploy activities which had social and cultural purposes?

All but two of the caregivers we interviewed identified themselves as average to avid readers of newspapers (at least several times a week), of books (finishing at least one a month and often weekly), and magazines (reading a weekly periodical). One mother's description was, '*this is a book house*'. Clearly, for the caregivers, reading served important personal needs.

Their personal uses reflect the practices of many families in New Zealand. Most New Zealand children see older family members reading newspapers, magazines, and books on a daily basis. Reading for personal pleasure and informational needs are still familiar and favourite activities of the adults in many New Zealand households, and do not appear to have declined substantially over a generation.[4]

> Reading for personal pleasure and information are still favourite activities of the adults in many New Zealand households, and do not appear to have declined over a generation.

Reading for religious purposes occurred in 10 families. The caregivers in these families had strong church links and Bible reading, and reading and singing of prayer and hymn books and other church material figured at least once a week and most times on a daily basis. Six of the seven Samoan families engaged in these activities. This is in keeping with many Samoan families in New Zealand where family devotions occur daily and where the church has an integrative function creating 'miniature Samoan communities' (Fairburn-Dunlop 1984). Two Maori and two Pakeha families engaged in these activities, too.

There were activities associated with reading and writing for another purpose, that of family cohesion. These mostly covered the sending and receiving of letters to extended family. All of the Samoan mothers said that they or someone in the household wrote at least one letter a week. Three of the Maori and three of the Pakeha families also sent and received letters for this purpose.

> One purpose of reading and writing activities was to maintain family cohesion.

All of the families read and wrote to maintain the family's domestic functioning. The activities ranged from those woven almost invisibly into the fabric of domestic life, such as the cutting out and writing of recipes and writing shopping lists, through to extremely significant family-threatening events. An example of the latter occurred in one family during the time the child we were following was a preschooler. The family was writing to the state-housing agency requesting an upgrading of the conditions of their (state-owned) accommodation.

Figure 2.2 An illustration of a close ambient activity captured in an early colonial painting by Joseph Merrett sometime during the period 1841–1843. It is captioned 'Maori woman reading', but could perhaps have the subtitle 'and a baby surrounded by an ambient literacy activity'. (from *Mrs Hobson's Album*, Locke & Paul 1990, Courtesy Alexander Turnbull Library)

> Community involvement constituted major forms of literacy activity, achieving significant social and cultural purposes.

In all of the families there was someone for whom writing (e.g. letters) was part of the work they carried out for a community or social agency. The formal responsibilities entailed a range of activities. Five of the Pakeha mothers (and a father in a sixth household) had served or were serving on preschool or school committees. One of the Samoan mothers had such a responsibility, too. Two of the Maori mothers held positions at their church which involved them in teaching or counselling duties, as did five of the Samoan parents.

This community involvement constituted major forms of literacy activity connecting the family with their communities, and achieving significant social and cultural purposes. Holding positions in church and school organisations has been common in Samoan households, reflecting the role of the church in maintaining Samoan culture and community values in New Zealand (Fairburn-Dunlop

1984). The two Maori families also belonged to churches which fostered strong family involvement (e.g. the Mormon church). As with the Pakeha families, their involvement reflects the practices of the church.[5]

In general these community forms of literacy and their significant purposes have tended to be invisible to, or at least their significance under-recognised by, researchers. Other forms of literacy, such as those relating to leisure and occupational needs, may have been valued more highly and been more obvious to researchers. Women have often had the roles in community organisations described in the SOL study, particularly in early-childhood education (McDonald 1970). The general 'invisibility' of their community work to researchers has spilled over to a lack of recognition of their literacy practices. Moreover, these uses of literacy in government, church, and other community organisations often have been in collaboration with others rather than achieving purposes through independent reading and writing. (Guthrie & Greaney 1991). Independent or personal ways of using literacy have been valued in school forms of literacy and hence collaborative forms are seen at best as relatively insignificant (Gee 1990).

In some sociocultural groups collaborative reading and writing activities occur more often than independent reading and writing (Heath 1983). Being more frequent activities of marginalised or 'at risk' groups, they have not been researched thoroughly as activities entailed in the socialisation of significant forms of expertise (see Heath, 1983, as an exception). The collaborative forms of literacy are, however, potentially important sources of ambient activities that provide channels for children's engagement.

Given the ways in which we selected the children for our study it is perhaps not surprising that among the ambient experiences were those created by older siblings. Books came home from school, and brothers and sisters read and wrote schoolwork in the presence of the children we were observing as 4-year-olds. In the junior classes of New Zealand primary schools taking books home to read to someone is a daily activity. From this time on, homework is a frequent experience in households throughout the levels of schooling, both in New Zealand (McNaughton 1992) and elsewhere (Stevenson & Lee 1990). One of its unintentional, but nevertheless significant functions, may be as an ambient model of literacy to young children not yet attending school.

These descriptions show family members engaged in many literacy activities, which served a number of purposes. Some of the more general of these purposes are summarised in Table 2.1 (p. 24). The categories of purposes come from a recent international review of literacy 'acts' by Guthrie and Greaney (1991). They describe the literacy acts in which people engage in terms of three general types: leisure acts, community acts, and occupational acts. These are not the same as activities, rather they are summaries of the general forms that adult reading and writing can take based on a number of research descriptions. Nevertheless, Guthrie and Greaney describe acts, like activities, as achieving a number of purposes or functions: personal empowerment, knowledge gain, participation in society, and occupational effectiveness.

Fitting the purposes of the different activities engaged in by the families in the SOL study into Guthrie and Greaney's categories is straightforward. Examples of

> Community forms of literacy and their significant purposes have tended to be under-recognised by researchers.

> Independent or personal ways of using literacy have been valued in school forms of literacy and collaborative forms are seen, at best, as relatively insignificant.

> Homework is a frequent experience in households at all levels of schooling. One of its unintentional, but nevertheless significant, functions may be as an ambient model of literacy to young children not yet at school.

purposes in the SOL study are described under the headings used by Guthrie & Greaney (1991). But as the acts can have multiple purposes, several of the examples could occur alongside each of the acts. For example, the leisure act of writing to relatives and reading their letters may serve purposes of personal empowerment, knowledge gain, and participation in society.

The value of this exercise of fitting the purposes identified in our study of 17 families into Guthrie & Greaney's categories is to establish that some purposes are shared across countries. In this sense there are some very general cultural messages. Our families had a range of literacy practices which they shared with families living in other schooled and industrialised communities, which fulfilled similar purposes. And like other families in these countries who work in so-called 'non-professional' occupations and have gained relatively few years of secondary schooling (the average for caregivers in the 17 families was less than three years), reading and writing at home specifically related to one's primary job in the workplace did not occur very often (Nash & Harker 1990; Guthrie & Greaney 1991). It mostly occurred with secondary occupations associated with positions in community agencies, such as being on school committees, and church occupations.

As with many other families, these 17 families selected literacy activities in which to engage, arranged for them to occur, and deployed them within the home. The sheer amount of time involved in these activities illustrates their potential as ambient socialisation experiences. A first answer to the question posed above (p. 21) is that families were engaged in activities which functioned to express and maintain their social and cultural lives.

Table 2.1 Examples of literacy purposes identified in 17 families in the SOL study, using the Literacy Act categories of Guthrie & Greaney (1991)

Literacy Acts	with multiple	Purposes/Functions
		1 personal empowerment
		e.g. personal pleasure
		e.g. religious needs
Leisure		2 knowledge gain
		e.g. educational roles
		e.g. church roles
Community		3 participation in society
		e.g. family cohesion
		e.g. daily living
Occupational		4 occupational effectiveness
		e.g. community agency role

But saying that activities are socially and culturally structured means more than noting that they achieve particular purposes for family members. It also means that the activities express and construct certain ideas and embody particular values both about literacy and about the family's identity.

This can be shown in a more specific example from the SOL study. It is the activity practised in the Samoan families of writing letters to other family members. This activity expresses and constructs beliefs about the nature of familiness. Samoan families in New Zealand have strong commitments to Fa'a Samoa (doing things the Samoan way). Fairburn-Dunlop (1984) describes a group of New Zealand Samoan parents' beliefs and commitments to Fa'a Samoa as including:

> ... loyalty to the family, village and nation [which] continued to dominate their behaviour and life in New Zealand, family commitments being maintained by frequent travel between the two countries, and indirectly by mail and money order remittances. (p. 104)

So Fa'a Samoa necessitates mechanisms for maintaining contact with members of the aiga (family). Writing is used as one of the tools and one of the language forms to achieve this. It also expresses beliefs about the role of literacy. If you like, the message is that this form of writing is a valued tool to be used to keep families in touch with each other.

> Writing letters may express beliefs about the role of literacy, e.g. a valued tool to be used to keep families in touch with each other.

Another example in these same families was Bible reading and family devotions (lotu). Several purposes are entailed in the activities involved in this practice. There are the purposes associated with religious beliefs, but there are also other purposes, including the cohesion of the family and the family commitment to shared beliefs, roles, and responsibilities in the Fa'a Samoa. For example, Tagoilelagi (1992) describing a family's lotu identified grandparents, parents, children, and grandchildren being present. During lotu the leader (who reads first) has a position of prestige. When the young 7- and 8-year-old grandchildren took over the role of leading the devotion this was seen on the one hand as fulfilling the older family members' responsibility for nurturing the younger ones (including aiding their reading development), and also as a way of telling the younger ones that they were cared for.

Joint activities

In the SOL study caregivers were asked also about the things they did directly with their preschoolers involving written language. There were four general sources of activities which they identified that someone in the household did regularly with their preschooler. These were reading books to them, teaching writing, library visits, and talking about print.

Book reading was perceived as a central activity. Although our interview questions only asked whether it occurred two to three or more times per week (or less), all of the families qualified their answers by saying book reading was a daily occurrence. This was a favourite child activity. The interview sheets record comments such as '... she hounds me', and '... he presents me with a heap of books and I have to work through them', and '... that's all she seems to do all day'. One of our research team recorded on an interview sheet a note that book reading was a daily occurrence. The word 'daily' has two exclamation marks beside it.

The overriding impression is one of insistent interest bordering on coercion in using anyone in the household who might be available for this activity. The implication, which was articulated in some of the interviews, is that children have the right to these literacy events. And the socialisation agents in the household often respond.

> Reading storybooks to children is a frequent activity in many homes. But there are differences between different groups of families.

Reading storybooks to 3- and 4-year-olds is a frequent activity in many homes in New Zealand (Phillips & McNaughton 1990) and elsewhere (Dickinson et al. 1992; Sulzby & Teale 1991). A generalisation from this research is that fewer families with income earners in 'non-professional' occupations and fewer families who are identified as part of a marginalised minority read to their children. And there are process as well as frequency differences, too, between families.

All 17 families read storybooks, but different texts were chosen. Some also read church books and the Bible (for example, during lotu) and other religious texts. It was an important feature of the Maori and Samoan households that during this activity more family members were present than in some of the Pakeha households during this activity. The records show that in these households there was an expectation for older siblings and relatives to take responsibilities in the reading.

Activities which involved the teaching of writing were strongly evident in all of the families. We called this teaching because all of the activities, as described by the caregivers, had explicit guidance or tutoring processes associated with them. The frequency, diversity, and socially mediated nature of the writing is impressive. But very little other data exists with which to compare these descriptions. Although it is known that many children explore features of writing before school, compared with storybook reading, international research into writing activities before school is 'meagre' (Sulzby & Teale 1991, p. 737).

> It is known that many children explore features of writing before school, but international research into writing activities before school is meagre.

Teaching the alphabet was as common as book reading, occurring in all families. We collected many instances of alphabet letters being written for children and them copying underneath. Similarly, the diaries which parents filled out recording the literacy events involving their 4-year-old refer to this common activity. The diary record for Colin at 4 years 11 months has the following entry: *Tried to write some letters from his name. Myself (present) ... helped.* The record for Jayne at 4 years 6 months notes, *Jayne wrote letters on the concrete outside with water and a brush and she asked me who's* [sic] *name started with each letter.*

The caregivers differed in how deliberate they claimed this tutelage was. But it was frequent, with identifiable patterns. Although it was often reported that the child would ask how to write letters, parental responses in the interview suggested a strong determination to have this learning happen. For example, it was a stated expectation that full letter identification should be acquired before 5 years (only three caregivers said they expected letters to be learned after going to school at the age of 5 years), and they reported deliberately engineering activities to ensure that this occurred, saying that someone had set out to teach letters to the child).

An additional letter-identification routine took place with the Samoan children. In all of the Samoan households the preschooler had been taught the letters using a singing sequence with a chart of the Samoan alphabet. It was reported that this happened at both home and at church.

While the teaching at home appeared to be very effective, it was not universally so. Two families reported difficulty in teaching the alphabet to their child. A Maori mother said that she had tried but felt she had failed; her child got '... hoha [bored] *about it and finds it difficult'*. A Samoan mother said that she had tried but her son wasn't really interested.

Children were also tutored in how to write their names. In fact, families generally believed that learning to write one's name should come before learning to identify the letters of the alphabet. All of the families expected their child to do this before going to school. Fourteen families reported that a model plus copying routine was used to teach children how to write their own name. Many of the families gave us examples, on pictures and all manner of scraps of paper (some examples are given in Chapter 3).

One of many examples from the SOL study of copying models of both the alphabet and a name is shown below. It comes from a child just before his fifth birthday. The model for the alphabet was written and he copied underneath. Only one letter shows a correction, to a lower case /r/. He could produce this more accurately than the upper case /R/. He attempted to write his name, too, from models not on the paper on which he was writing. Twice his attempts involved the reversal of letters and a right to left sequence. He corrected his third attempt, which began with a right to left sequence and finally produced an accurate left to right rendition.

Sample 2.1 The alphabet and his name copied from models, by Jeremy, aged 4 years 11 months

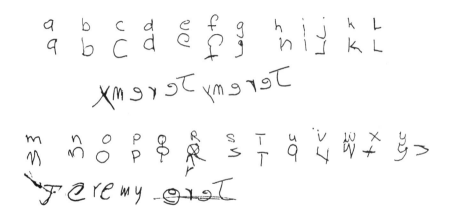

Caregivers develop expectations about the literacy development of their children, just as they do in other child-rearing tasks, e.g. learning to write their names before starting school, and writing all the letters of the alphabet as the next achievement at school.

Clearly, caregivers develop expectations about the literacy development of their children, just as they do in other child-rearing tasks. Bruce McMillan (1983) repeated an Australian questionnaire with a group of 45 mothers involved in the New Zealand early childhood organisation of Playcentre. The questionnaire asked about when the mothers expected their children, who were between 4 years 6 months and 5 years old, would achieve certain developmental tasks. In general, the mothers expected that they would be learning how to write their names at the time of the interview. The mothers expected the writing of all the letters of the alphabet as the next developmental task, when their children went to school.

McMillan compared his results with the original Australian data. The timing of this developmental task, knowing how to write one's name, was significantly different from a sample of Australian-born mothers, who in turn had significantly different expectations from those of a group of Lebanese-born Australian mothers. There were differences in the timing of expectations between the middle-class mothers in McMillan's study and the caregivers in the SOL study, which are related to social and cultural values. For example, all but three of the mothers expected letters to be learned before the child went to school.

Writing sentences also occurred. The description from parents revealed an activity very like that which takes place in junior classrooms in New Zealand using 'dictated stories' and 'caption books'. The child gave a sentence and someone wrote it down for the child to copy. Nine families did this, five of whom were Pakeha. It seemed a very obvious activity, strongly promoted in Pakeha households. Two Maori and two Samoan families also said that their children copied sentences. Variations to this activity included writing greeting and Christmas cards.

Library trips were major regular joint activities taking place weekly or every two weeks at designated times. This was especially so for the Samoan and Maori families. Only two of the Pakeha families said that they went regularly.

A final activity deserves mention. It occurred in all the households as reported in the parent diaries. It is the least structured of all the activities. It could be called 'talking about ambient print'. With apologies to Frank Smith (1978), who established this activity in the consciousness of reading theorists, a subtitle might be 'the old cereal box example'. Previous writers, particularly when describing children who read early, have referred to children's interest in print around them. Signs and labels and print arouse interest (Durkin 1966).

Reports such as the following appeared in the diaries we asked the caregivers to keep: *Went shopping at 3 Guys. K ... asked what certain items were and told me what others were; ... talks about pictures on cereal box. Wants to know what words say.* A child was reported as saying, *... I can read the words on the milk bottle.*

One telling example occurs in the diary records kept by Mat's mother. The full record is reproduced on page 29, and the focus of the activity is shown below. This mother concentrated on recording conversations relating to written language, and often these occurred at breakfast. Her record for one breakfast reads, *Me and Mat have breakfast Mat asked what this say on the salt container I said, it said see how it runs. And Mat asked Mum can the salt run.*

Figure 2.4 The focus for Mat's question — 'Mum, can the salt run?'

Sample 2.2 A joint-activity diary recorded by Mat's mother

Reading and Writing
Please fill out this sheet three or four times during the day for two days.
Day 2

| About what time did the reading or writing take place, and for how long? | What was the reading or writing that took place? | Who was there and what were they doing? |

> 7.00 am. I filled this paper and Mat. said to me mum I want to go to the university.
> 11.00 am. I read the Auckland Star. Mat. saw a picture of football men and he said that's Bop and Joe Stanley.

> 8.15 a.m. Mat asked me mum that the supa bread for breakfast.
> 2.00 pm. Mat. watch the TV. he asked me where are those picture in the TV. come from.

> 8.00 a.m me and Mat. have breakfast Mat. asked what this say on the salt container I said, it said see how it runs. and Mat. asked mum can the salt run.

> 7.00 am. me and Mat have porridge for breakfast mat. asked where are the shops get all those food from.
> 2.30 p.m. Mat watch the TV. he asked me are all those picture on that TV are real. can the robot walk on the ladder.

As with the observations about ambient print there is much evidence to show that these families arranged for joint activities to occur; that they deployed particular resources to enable them to occur, and that they selected particular activities in which to engage the child. When their child started an activity and needed them, or approached asking a question, these families often harnessed their physical and social resources to focus on the child's growing expertise.

Personal activities

The interviews and diaries from the SOL study show that parents were aware, or became aware when they started a diary, of considerable amounts of time spent in personal reading and writing. This increased sensitivity to everyday activities is a feature of research in families which I continue to discuss in later chapters. It comes about because attention becomes focused systematically on what children are doing. Previously, their activities have been so ordinary that they are (literally) unremarkable. In the diary, the caregivers were asked to record three things: *when* anything involving reading or writing took place; *what form* that reading and writing took; and *who else* was present.

The diaries show that personal activities often occurred and took many forms. One mother's diary over two days recorded activities when the child was by herself and when her mother, her school-age sister, her school-age cousin and her nana also were close by. In the 300 minutes recorded approximately 100 minutes in total were recorded when the child was engaged in activities involving reading and writing carried out entirely by herself. Sixty minutes of this time involved children's educational television. Appropriately, programmes such as *Sesame Street* and *Playschool* were recorded in the diary because of the written language which is displayed in the programmes.

Many of the recorded activities indicate independent play and experimentation in close proximity to others. For example, for 10 minutes on one of the days the same child drew in her sister's old schoolbook, copying her sister's work. Her sister and her mother were nearby.

> Personal activities depend on family resources: they are reflections of what children have observed and done with other family members; they require space, as well as time and a setting; and they require material resources.

These personal activities are also dependent on family resources. Firstly, because they are reflections of what children have observed and the sorts of things children have done together with other family members. The diary record of this child's activities shows her doing things that are reconstructions and generalisations of things she has seen others do, and things that she has participated in with others. But secondly, this play, and practice and experimentation need direct resourcing, too. These activities require space, as well as time and a setting. And they require material resources — things to write on and with, and the availability of things to read.

Activities: cultural and social voices

The patterning of activities within the practices of 17 families is explained in terms of their social and cultural identities. That is, their activities both reflected and constructed particular ideas and values. But they were not dormant channels for the direct transmission of concepts held in common by families who have similar social and cultural identities. Moreover, their ideas and values were not static or fixed as they reared their children. I discuss in chapters 9 and 10 how parents may construct knowledge from formal and informal educational experiences.

There are two principles here. One is that activities are socially and culturally based. The other is that the messages to do with identity are dynamic. Both of these principles about the nature of culture and society provide a means of understanding why families in general, and these families in particular, select, organise and deploy particular sorts of activities.

> Caregivers, particularly in multilingual, multicultural, industrialised countries encounter a collective culture in which messages of varying ambiguity and redundancy exist.

Caregivers, particularly in multilingual, multicultural, industrialised countries such as New Zealand, encounter a collective culture in which a variety of messages of varying ambiguity and redundancy exists. Multiple cultural meanings and messages exist in socialisation processes (Valsiner 1994a). Parents, and teachers too, are active agents in their socialisation roles (Goodnow & Collins 1990). They select from and adapt to these multiple heterogenous 'voices' which are available to them. They reconstruct and reinvent their actions of child rearing in relationship to cultural messages.[6]

Although in general the 17 families were like many other families in New Zealand, there were both differences between families in the groups and differences with other research descriptions of families who share similar cultural identities, educational levels, and occupations. For example, Nash (1992) reports fewer Pacific Island women were described by their daughters as reading recreationally on a daily basis than non-Pacific Island women. In the SOL study there were Samoan mothers of 4-year-old daughters who often read; one mother reported that she read *'Novels. Anything [I] can get hold of and information [I] read at work.'* In international comparisons, families who have had relatively few years of schooling and whose income earners are not in professional occupations less frequently engage in storybook reading with their children (Dickinson *et al.* 1992). This is particularly true of African-American working-class families in the United States (Heath 1983).

The ways in which the 17 families were similar to, or different from, other families in New Zealand reflects their active 'construction' of social and cultural messages for their family system. The processes of constructing ideas of socialisation can be illustrated in a very typical joint activity — reading storybooks to children.

> In international comparisons, families who have had relatively few years of schooling and whose income earners are not in professional occupations less frequently engage in storybook reading with their children.

The meanings of who reads to preschoolers

Reading to preschoolers is widely valued in New Zealand. The value is detectable in many messages available for parents and is legitimised in much of the official advice given to parents. For example, a book for keeping developmental and health records is given to mothers of all children born or registered at maternity hospitals by the Department of Health. This Health and Development Record (the most recent version is 1987) provides for record keeping related to visits by public health nurses and the network of neonatal and infancy nurses (Plunket nurses) which has operated for several decades seeing most new babies and their families. One of the developmental tasks relating to the development of sight is provided at 18 months by a health nurse or doctor who checks whether the child points to pictures in books. Among the items of advice given to parents for preschool development is reference to reading to children; parents are urged to *'Read them stories'*.

> Reading to preschoolers is widely valued in New Zealand.

In three studies of 3- and 4-year-olds we have found that most families read to their children on most days of the week. The diaries of 10 middle-class Pakeha families over 28 days revealed that reading took place on six days out of seven, with an average of 22 books read over a week (Phillips & McNaughton 1990). While we did not collect information in the same way in our study of the 17 families, the interviews yielded the result noted earlier. Again, it was found that reading was close to a daily occurrence.

Despite social and cultural differences between families in these two studies both groups of parents had been chosen on criteria relating to the children's likely success in school forms of literacy. In the SOL study this was judged from the success of older siblings at school. In the study of middle-class Pakeha families this was done through descriptions of families' literacy practices. A third study involved a group of eight Tongan families (McNaughton, Ka'ai & Wolfgramm

1993). These families had been selected from members of a local church and the likely success of their children at school was not considered in their selection. On other criteria they were most like the 17 families. The mothers had completed three years of secondary schooling and the fathers, on average, had completed just primary schooling. Their diary records over a week reveal book reading (including family devotions) on five days of the week. Over the week, they read an average of seven different books to the children.

I explore cultural bases of book reading more extensively in Chapter 6. But the presence in literacy activities of social and cultural values can be shown in one seemingly simple characteristic of book reading that was different for different families in the three studies. It is the characteristic of who was present during book-reading sessions. Technically this is an analysis of differences in 'participation structure' in speech events. Such differences relating to how children learn have been explored with different cultural groups in classrooms and in family settings (Cazden 1988; Schieffelin & Ochs 1986). Who participates, as well as how many participate, in an activity reflects socialisation values. So much so that under some conditions the effectiveness of classrooms as settings for learning is determined by how well the participation structures match those that children are familiar with in their home culture (Tharp & Gallimore 1988).

Important views of family roles and the nature of literacy are carried in the participation structures of book-reading sessions. All of the sessions of book reading taped with the Tongan families were triadic or multiparty (involving more than one other person) rather than dyadic (involving two participants). All of these families had two or more children so this finding may not be surprising. But the list of participants includes more than siblings who were present; it included particularly, aunties and cousins. By contrast, even though seven of the 10 Pakeha middle-class families in the first study had two or more children, only a third of the sessions recorded over a 28-day period involved more than one reader and the preschooler. The 17 families were in between. But in all of the Maori and Samoan families there were examples of multiparty sessions while in only two of the Pakeha families was this the case. This was despite the fact that in all of the families there was an older sibling.

Group learning is a preferred pedagogical mode for Maori (Smith 1987). This preference does not exclude personalised learning interactions, which can be seen occurring in group settings and shared activities. But the development of individual expertise carries responsibilities for the group. The preference is derived from significant cultural values associated with the concept of whanaungatanga (literally, familiness, in terms of the extended family). One expression of this principle in the socialisation of Maori families is the relationship referred to as tuakana-teina (Nepe 1990). This describes the responsibility of the older sibling or more expert member of the group to take responsibility for the needs of the younger or less expert member of the group.

These values can be seen operating in specially constructed Maori settings. Te Kohanga Reo (Maori-immersion preschools, literally 'language nests') were deliberately set up in 1982 to reflect Maori kaupapa (principles), one of which is whanaungatanga (McNaughton & Ka'ai 1990; Hohepa et al. 1992). Descriptions

> Socialisation values are reflected in who participates in an activity, as well as how many participate. So much so that under some conditions the effectiveness of classrooms as settings for learning is determined by how well the participation structures match those that children are familiar with in their home culture.

> Group learning is a preferred pedagogical mode for Maori. But this preference does not exclude personalised learning interactions.

of interactions in Te Kohanga Reo highlight the significance of multiparty contexts for language learning. They also show the presence of tuakana-teina relationships, expressed in interactions between peers and between children and teachers (see also p. 116).

> Maori socialisation patterns and some Pacific Islands patterns identify older siblings and extended family as often being expected to take immediate responsibility for the needs of the younger siblings.

Similarly, descriptions of Samoan socialisation patterns identify older siblings and extended family as often being expected to take immediate responsibility for looking after the needs of the younger siblings. The responsiveness of older siblings and extended family to young children, at times being present instead of a parent, carries meanings about differential status of child rearers. Similarly, child-rearing patterns reflect values associated with the priority of familiness, including loyalty to the extended family unit (Ochs 1982). The strength of this value can be detected in Samoan community values for school aims. For example, Samoan families do not rate the goal of 'teaching children to think for themselves' as highly as do Pakeha families (McNaughton et al. 1992a).

In Tongan families similar principles are at work as regards who reads to the preschooler. In traditional society, as soon as a child was weaned and became less reliant on his or her mother, older children in the family took charge. The values of this responsibility within the group are derived from the principle of Fatongia (the Tongan way). Older siblings have a role to care for the younger (Wolfgramm 1991).

The similarities and differences between families on this one dimension of participation in the activity reflect caregivers' views about socialisation. In general, caregivers had constructed from the messages available to them beliefs about the importance of reading to children. Yet within this they had constructed ways of participating that reflected specific beliefs, values, attitudes, and knowledge.

Summary

This chapter has focused on the role of the family system in socialising children into ways of using written language. Two propositions have been explored in detail:

> Families socialise children into their literacy practices, which reflect and build social and cultural identities.

Proposition One: Families arrange time and provide resources which socialise children into their practices of literacy.

Proposition Two: Family practices reflect and build social and cultural identities.

A third proposition has been introduced here. This one, which is examined in detail in Chapter 4, states that:

> Literacy practices include goals, rules for participation, and ways of carrying out the activities.

Proposition Three: Literacy practices are expressed in specific activities which have identifiable constituents. These include goals, rules for participation, and ways of carrying out the activities.

The next chapter is devoted to an analysis of how development and learning take place, firstly by concentrating on the child's role.

Implications

For families, educators, and researchers

> Everyday activities involving written language are important sources of learning.

Families: Those everyday activities involving written language in which all members of the family engage, are important sources of learning for children. This means that frequent opportunities should occur for indirect participation, as in observing, 'playing' with, or exploring the forms of written language that are valued by the family, and for direct participation, as in reading books together.

> Family practices are dependent on access to educational resources.

Educators: Family practices are dependent on access to, and participation in, educational resources. These include resources such as local libraries, early-childhood settings, and churches. It includes also the community resources such as communication and transport which facilitate participation. Depletion or competitive use of such resources puts family literacy practices at risk.

Researchers: The research agenda includes understanding families' literacy practices and showing how these are dependent on social and cultural resources (including economic, political, and community resources). In turn, this depends on developing ways of collaborative research in which literacy practices are represented adequately in terms of the meanings they have for the family.

Further Reading

The family as a system and parents' ideas in socialisation:

Bronfenbrenner, U. (1986). 'Ecology of the family as a context for human development'. *Developmental Psychology*, 22, 723-742.

Goodnow, J. J. & Collins, W. A. (1990). *Development According to Parents: The Nature, Sources, and Consequences of Parents' Ideas.* Lawrence Erlbaum, Hillsdale, NJ.

Valsiner, J. (1994a). 'Culture and human development: A co-constructivist perspective'. In P. Van Geert & L. Mos (eds.). *Annals of Theoretical Psychology.* Vol. X. Plenum, New York.

Family practices of literacy:

Heath, S. B. (1983). *Ways with Words: Language, Life and Work in Communities and Classrooms.* Cambridge University Press, Cambridge.

Sulzby, E. & Teale, W. (1991). 'Emergent literacy'. In Pearson, P. D., Barr, R., Kamil, M. L. & Mosenthal, P. (eds.). *Handbook of Reading Research, Vol 2.* Longman, New York.

Wolf, S. A. & Heath, S. B. (1992). *The Braid of Literature: Children's Worlds of Reading.* Harvard University Press, Cambridge, MS.

End of chapter notes

1. I have followed Goodnow and Collins' (1990) use of ideas as a general term which refers to meanings and systems of beliefs. But it also carries the notion of linkage with action and so includes expectations and attitudes. Goodnow and Collins also associate the term 'idea' with affect and feeling. These affective features of ideas are highlighted in the present text by the reference to values.

2. Early colonists or immigrants, who were mostly British, were identified by Maori as Pakeha. Literally this originally referred to being foreign but it has came to be used by some families to self-identify as having descended from Anglo/European forebears.

Yensen, Hague & McCreanor (1989) define Pakeha as a general cultural group denoting 'the original British settlers, their descendants, and subsequent immigrants of mainly European descent' (p. 14). I have retained this term of cultural identity as carrying important meanings within the bicultural context of New Zealand, although not all families would use this term preferring to be called European/English/Anglo/ or white New Zealander. Terms used for research purposes in New Zealand include all of these.

3. This focus on families who, on some general criterion such as SES, might be 'at risk', but whose children were achieving well at school, has been adopted relatively infrequently. One notable exception is the study by Snow *et al.* 1991. Their reasons were similar to ours. General criteria such as SES and ethnicity are not psychological mechanisms; at best they provide what Bronfenbrenner (1979) calls a 'letterbox' index for processes. It is the processes at work behind the letterboxes that need analysis. Children from low-income and 'minority'-group families are often problematised by research models and few research resources are committed to examining robust psychological and educational processes in their lives. Furthermore, the mechanisms of 'success' for low-income families may be different from the mechanisms of success for other groups (see Snow *et al.* 1991).

4. In 1982 Guthrie (1982) summarised earlier surveys of reading habits of New Zealand adults. These surveys showed that in international terms high levels of reading occurred in New Zealand families. A survey of adults conducted in 1989 ('Life in New Zealand', Ansley 1990) asked 10,800 New Zealanders aged between 15 years and older about leisure pursuits. Reading was by far the most popular leisure activity for women in their 40s and 60s; about two-thirds listed it in their five favourite leisure activities. For men in these age ranges it was on average the second most popular leisure activity.

Descriptions of family reading habits also come from the self-report of the 3000 9- and 13-year-olds in the most recent IEA (International Association for the Evaluation of Educational Achievement, Wagemaker 1991) studies of literacy. Three-quarters of the children reported that their homes received a daily newspaper and 40 per cent of them reported that there were 200 or more books at home.

An interesting issue arises in these data, however. The recent IEA studies (Elley 1992) show that in New Zealand, television watching has increased dramatically over the last 20 years for 9- and 13-year-olds. Reading for pleasure and other personal functions involves activities into which one needs to be socialised (either directly of indirectly). If this is so, the 9- to 13-year-olds watching high rates of television in 1989, when the IEA research was carried out, may provide models of reading as ambient activities less frequently in their families when they become parents. The effects of increased television viewing may be felt across generations.

5. In other accounts of Maori families literacy activities stem from participation in hapu (extended family) and iwi (tribal) organisations. Literacy activities to promote hapu and iwi needs have been a feature of Maori literacy from the immediate post-colonial contact times. For example, newspapers figured in the early activity of tribes to develop critical community awareness and links among communities to resist colonial threats to sovereignty (Sinclair 1991). This has contemporary equivalents. Early political goals are mirrored in the significant reading and writing activities undertaken by hapu, iwi, and other Maori groups to achieve, among other things, educational and economic goals (e.g. Murray 1974).

6. A number of writers have developed Bakhtin's (1981) concept of heteroglossia or multiple voices in language. Valsiner (1994a) has developed an argument about the essential heterogeneity of cultural meanings and messages that exist in socialisation processes. Both socialisation agents and children encounter an organised social world in which a variety of messages of varying ambiguity and redundancy exist. These features are incorporated into Valsiner's general concept of collective culture (see also Cazden 1993a; McNaughton 1994a; Wertsch 1991).

Part Two

Actions and Interactions

Chapter Three

What's in a name?

> **Focus**
>
> **The child's role in learning and development**
> - Systems for learning and development are constructed within activities.
> - These systems are a product of the child's actions and the actions of significant others.
> - Descriptions of a child learning to write his name are used to show how children's actions contribute to their learning and development.

Literacy activities provide a framework for the development of expertise. Teaching and learning mechanisms form within this framework.

Literacy activities provide a framework for the development of expertise. The socialisation model (p. 3) proposes that within this framework teaching and learning mechanisms form which provide the vehicle for development to take place. In this chapter and the next the theoretical descriptions begun in Chapter 2 are extended as I examine the immediate mechanisms which enable expertise in written language to be co-constructed. In the present chapter the role of the child occupies the foreground, while in the next chapter the role of social interactional mechanisms becomes the focus.

The analysis starts from a case study. It describes, in part, a child's developing skill in being able to write his name. The case is personal, involving my younger son. The action takes place over 21 months from before his second birthday to after his third. Before two he was unable to write his name. By the time he was 33 months old he was able to write his name unaided.

Guiding this case study is a question. What went into this 'situated' expertise? To answer this I focus on one part of his expertise, in particular his control over the first letter /H/. The burgeoning of control is analysed in terms of how he acted to overcome the 'problem' of writing his name, building his expertise in the process. The description serves to highlight the social interactional vehicles which provide a basis for his mediated solutions. These are described as systems for learning and development. Describing these processes of co-construction enables us to answer the question of what went into writing his name. This chapter analyses what Harry put into the construction of this expertise. The more specific analysis of our joint productions occurs in the next chapter.

Developing Expertise as a Learner

Samples of Harry's attempts to write his name were collected as they occurred. The method of collection could not be called systematic in the structured sense of happening at specific regular times using predetermined modes of collection. Writing at home, at least in our home and in those others in which researchers have attempted to collect samples (Heath 1983; Sulzby & Teale 1991), does not take place at regular times. In keeping with an intention to analyse these everyday episodes, occasions which might be deliberately structured to probe his (or our) attempts were eschewed. Attempts I observed taking place and in which I was involved were collected systematically. That is, I made detailed notes at the time of collection. Similarly, I collected details from others who might have been present when I collected samples from episodes in which I had not been present.[1]

Samples, including those in which I was directly involved, were collected in a variety of settings. The fact that a number of settings were involved illustrates how learning to write out of school is triggered by many needs. Among the settings represented in the samples that are shown on the following pages are his home, his grandparents' home, cubicles in banks, and his car seat.

> That a number of settings may be involved in a young child's writing activity illustrates how learning to write out of school is triggered by many needs.

Parts and wholes

The first samples come from 4 August 1989. Harry and I were at home together. He was playing in his bedroom with felt-tip pens and blank paper. He asked me to write 'Harry'. The resulting sample which he constructed from my model shows he knew several things. The Piagetian developmentalist Ferreiro (1985) provides a useful set of descriptions of this developing knowledge base. Her analysis suggests that while Harry was working on the general problem of how to represent his name, he had to solve a multitude of attendant problems. One example is the relationship between the whole word and its constituent parts (/H/ /A/ /R/ /R/ /Y/).

From the 4 August samples the inference can be drawn that already Harry knew that writing employs non-iconic symbols (he was not trying to draw 'Harry', he was trying to write 'Harry'). Also, he knew that the symbols have properties. He was aware of something Ferreiro calls the hypothesis of 'minimum quantity'. That is, he was performing on the basis of a guiding hypothesis about the nature of writing, that there is a fixed length to words. His attempt to write Harry was roughly the length of his name as I had written it (mine is the one on

> While a child works on the general problem of how to represent their name, they have to solve attendant problems. One is that writing employs non-iconic symbols, and these symbols have properties.

the top left!). Clearly he was aware of components. The most significant of these being the beginning /H/.

On 4 August Harry had attempted /H/ (top right of Sample 3.1). The sample 6 days later (10 August) shows him still trying to gain control over the vertical strokes and the horizontal stroke. He was writing at the breakfast table by himself. He used the blank back of a circular to members of the Board of Trustees of the local primary school that had been left on the table (both of his parents were elected community members on governing boards), and he used a nearby pen. His first movement with the pen was a full vertical stroke. The performance of the horizontal feature shows the presence of an approximation strategy, using control over shapes he already had (the circular features of his /a/ are present in his performance of the horizontal feature of /H/). It also shows his growing awareness of what the critical or defining features might be. At the same session Harry again tried the /H/ by itself several times as well as his whole name.

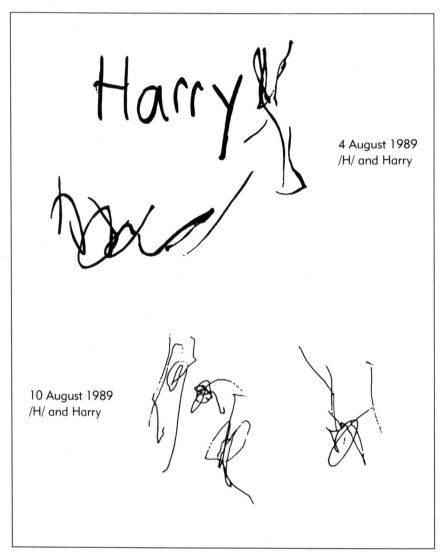

Sample 3.1 The development of writing a name: Samples of Harry's writing: /H/ and Harry

4 August 1989
/H/ and Harry

10 August 1989
/H/ and Harry

Sample 3.2 The development of writing a name: more samples of Harry's writing

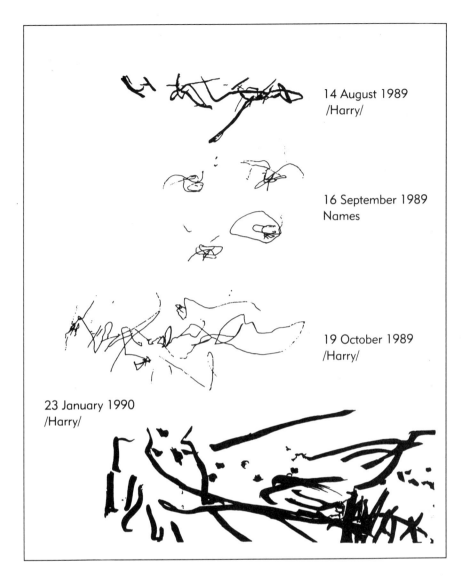

The sample seen four days later (14 August, above) was written on the back of the draft of a book review being prepared by his older brother. The full front page is not shown but the words written by his brother had been circled with a felt-tip pen. His attempt at Harry shows a passable /H/. Control over the horizontal stroke had developed.

Harry's growing knowledge of words is shown in a sample recorded one month later (16 September, above). He attempted to write the names of his family. As he wrote the four 'names' shown in the sample he said they were his two grandparents and two other members of the family. I am represented by the bigger circular rendition on the bottom right. While I described his knowledge as including the idea that writing used non-iconic symbols, it might be nice to imagine my status being represented here. The humbling truth is that it was probably a spontaneous serendipitous flourish.

The 16 September sample is included also to show that the knowledge base

> A knowledge base is changeable and adaptable.

Ferreiro describes is changeable and adaptable. Harry sometimes held concepts, sometimes not. He knew names have different constituent parts and he was able to represent different parts in his attempts at Harry. But it was beyond his representational capabilities to use this hypothesis of 'minimum quantity'. So rather than do this he expressed the constituent differences in the different names in another way, thus maintaining the hypothesis. His solution was four wholistic representations, a regression to his productions of six weeks earlier.

But this shows the development of a further concept. It is a primitive form of the concept Ferreiro terms the principle of 'internal variation'. This principle applied to different words holds that the same elements can't say different things. Or that different words have to be represented by different elements. Harry couldn't write different elements but he certainly attempted to distinguish what otherwise would be words that might look as though they have the same elements.

> The principle of 'internal variation' applied to different words holds that the same elements can't say different things.

A flourishing of constituent elements

October found Harry continuing to develop his full name using a strategy of inventive practice with constituent elements. The name is a mass of semicircular flourishes and lines. The performance of /H/ regressed, which is a feature of development in other areas when development involves learning how to control and embed skills in increasingly more general forms. Control of this element lessened as he attempted to gain control of the other elements. The writing was on the back of a blank envelope taken from my desk. He wrote in the space for writing the sender's name and address, showing that his knowledge was situated in particular activities. (See Sample 3.2, p. 41.)

> Regression in the performance of a single letter is a feature of development in other areas when development involves learning how to control and embed skills in increasingly more general forms. Control of the initial element lessens as the child attempts to gain control of the other elements.

This inventiveness and experimentation continued into the new year. After dinner on 23 January 1990 Harry sat down at the table and asked for paper and 'tip' (felt-tip pen). As he wrote by himself (his parents were talking together nearby) he said out loud, punctuating each letter with a pause, 'H-a-r-r-y'. The function of his language at this point is very significant. As early as 9 August he was saying 'H for Harry' repeatedly as he attempted to write /H/. On 19 October, in response to my question about what he had written, his reply was 'H-a-r-r-y', again punctuating the identification of each letter with a pause. The sample from 23 January, on page 41, is the first record of Harry saying each letter as he attempted to write it. I discuss where this identification of elements came from in the next chapter.

Gaining control

Harry's performance of his name during this time of the flourishing of internal constituents was full of vertical and horizontal lines. Some were straight and some were curved. A month later (14 February, Sample 3.3, p. 43) and Harry's expansive period was over. His attempts were contracted and economical. The latter feature might have been influenced by the setting for this sample. He wrote on deposit slips at the bank while I was involved in transactions with a teller. He wrote on a line following the procedure I had used. Again, aloud in the bank, he identified each of the letters as he wrote them,

'*H-r-r*', missing out the /a/ and the /y/. On this and on a second deposit slip he was using his knowledge of one to one correspondence, one sign for the identification of one letter-sound. On this second deposit slip (not shown here), he added to the '*H-r-r*' by saying '*Pert*' and created a sign. He did not identify the elements of this middle name (Pert), nor of his surname, McNaughton, when he said it and made a sign for it as well. Nevertheless, his identification and representation of chunks of speech shows the process of generalising his knowledge of correspondence.

This five-month period, from September 1989 to February 1990, did not involve major shifts in knowledge. It involved increasing control over the medium, the means of representing his name. Perhaps following Ferreiro we could call this an assimilatory phase, meaning a period of consolidation and learning to control under different conditions.

By 18 March 1990 (Sample 3.3) Harry was back in full control of the /H/. Two

> A child's identification and representation of chunks of speech shows the process of generalising their knowledge of correspondence.

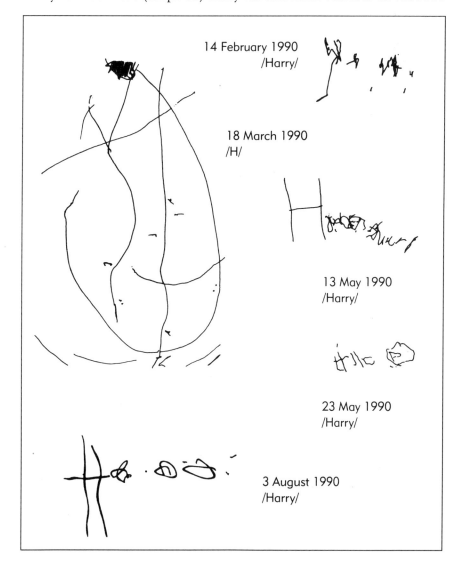

Sample 3.3 The development of writing a name: more samples of Harry's writing

nearly parallel vertical lines with a nearly horizontal line can be seen in the sample on that day. It was written on yet another piece of bank stationery. He said out loud, '*straight down then across for Harry*', paraphrasing something I had said to him two weeks before (see Chapter 4, p. 60).

Two months later (13 May) Harry was writing his name with a perfect /H/ and what Clay (1975) describes as a 'gross approximation' of the following letter string. It was written on a card for 'Mother's Day' and it was Harry's attempt to sign his name. (See Sample 3.3, p. 43.)

Solutions

By May (see samples 23 May and 3 August, p. 43) Harry had achieved an important production reflecting a major shift in his knowledge. He attempted a one to one representation of the elements in his name. His solution of the part/whole dilemma was in its final stage. All he needed from this point was to gain control of the other elements in his name. This was completed by the early part of the following year (May and July 1991, Sample 3.4).

Sample 3.4 The development of writing a name: a final sample of Harry's writing

May and July 1991

The meaning of expertise

This small slice of Harry's developing expertise shows his contribution to its construction. We can see what he was working on as well as the results of his problem solving. The samples are both instances of the products of his work and also instances of the processes he was employing to become more expert.

Learning occurs as children like Harry act on the world around them. The expertise that develops from that learning has certain properties; partly these are captured in terms of three components of skilful performance (see McNaughton 1987). Becoming skilled includes coming to know about how, when and where to use one's skill. It also includes the development of strategic and flexible ways of performing the skill. A third very important component of becoming skilled is the development of self regulation, the capability to engage in checking, examining, and modifying one's performance. Each of these components is reliant on the others; they are interdependent. For example, the more

> Learning occurs as children act on the world around them. The expertise that develops from that learning has certain properties, including skilful performance.

one knows about being skilled the more one can see perturbations in one's performance and can learn from overcoming those problems (such as being inaccurate or incomplete).

Adding the concept of expertise to the idea of being skilled enriches it in several ways (see Wood 1988). The idea of an expert suggests a person who has been trained and practised in the ways of employing the skill. At one time the learner was a novice who then develops towards someone's or some group's view of being fully expert. Barbara Rogoff captures this notion in her expression of 'increasing participation in communities of practice' (Rogoff 1990). As suggested in Rogoff's expression, expertise also is associated with an emphasis on development in particular contexts of use. A person is an expert in something, able to do something in specific situations. Using the term 'expert' enables us to avoid some of the more mechanical connotations of 'skill' which are associated with everyday, non-technical uses of the term.

> Expertise also is associated with an emphasis on development in particular contexts of use.

Knowledge and performance

One component of developing skill is a knowledge base. In the course of coming to write his name Harry acquired knowledge about writing (see Chapter 4). That knowledge included different aspects of how one can write one's name (including where you could write your name as well as how you can write a name), the functions of writing (such as someone being pleased to get a letter from you), where you can find writing, and knowledge about how you hold and control your fingers and co-ordinate that control with attention when writing.

> One component of developing skill is a knowledge base, e.g. in writing, the knowledge is about the functions of writing, where writing is found, and how to hold and control one's fingers and co-ordinate that control with attention.

Associated with this knowledge were ways of producing the writing. Several performance strategies have been identified in the sequence. What Harry was learning were ways of solving the task of performing as a writer (of his name). His performances were strategic. That is, he varied them according to the circumstances of what he was writing on and where he was writing.

The strategies themselves developed. Harry developed a plan for starting his name (*'H for Harry'*). There were also strategies for producing the /H/ with which his name began. And there were strategies for producing the constituent elements of HARRY. In part, all of these strategies involved formal and explicit algorithms, that is rules to guide the productions. One was a verbal algorithm for producing an /H/: *'Straight down then across for Harry.'* By saying *'straight down then across … '* he was able to produce the general model of an /H/ under many different circumstances.

> Algorithms can be developed for the segmentation of the whole word, the identification of its elements, and producing the right sequence. Algorithms may involve graphemes, and their associated phonemes.

Other algorithms developed for the segmentation of the whole word, the identification of its elements, and producing the right sequence. Sometimes these involved the identification and labelling of **graphemes** (the letter or alphabet symbols which represent sounds in the oral language) and sometimes the production of **phonemes** (the individual sounds) associated with the graphemes. These occurred when he broke down his name into constituent elements, either by naming the letters (H-a-r-r y) or sometimes producing the sound (phonemic) components (Huh-ah-rr-ee).

At different stages strategies for representing the whole of his name using approximations for these constituents was employed. This enabled him to produce an approximation of his whole name without yet having the skill to produce a fully expert rendition of ARRY (for example, his productions in August 1989, p. 41).

Acts of learning

Harry was actively learning through these performances. Developmental psychologists describe children as inherently active, talking about them as problem solvers, as engaged in actions of checking and matching and regulating (see Chapter 1). This idea of the active child is a legacy of a number of theoretical traditions in psychology, but most obviously Piaget's theoretical work and some information-processing approaches to cognition (Wood 1988).

> Children are inherently active — problem solvers engaged in checking, matching, and regulating.

A child has inherent capabilities for making sense out of their world which develop as they become more expert in their lives. Developmental psychologist Bruner (1983) called these basic learning mechanisms cognitive endowments. They enable the child to detect patterns and regularities during activities, and provide the child with a readiness to search for connections between actions and consequences. The endowments mean that learners have abilities to match patterns and imitate (see Chapter 4). These basic learning mechanisms are primitive at birth. They develop and are deployed in ways that are specific to the context of use and become increasingly strategic and self-controlled over time. Chapter 4 shows how they can be activated during activities.

> A child's inherent capabilities for making sense out of the world develop as they become more expert.

> Learners have basic learning mechanisms. These develop and are deployed in specific ways, and become increasingly strategic and self-controlled.

Some of Harry's learning activities were quite obvious in his performance strategies which were described earlier. Harry employed an explicit learning strategy for matching patterns. The sample from 4 August (p. 40) was initiated when he asked me to write 'Harry' for him, which he then copied. He could match my pattern. This strategy was generative and generalisable. On 23 May 1990 Harry asked me, *'Can you write Talia?'* (his sister's name). He knew that models existed for names, and that the strategy of matching patterns was transportable. In August of that year he asked how to do an 'S' for Sam (his older brother). Also in August he asked how to *'do childcare'*. This produced an interaction which illustrates joint activity at a very detailed level and is described fully in the next chapter.

Another explicit learning strategy was to actively search for and detect recurring patterns. Having found a pattern, Harry was also likely to check it out with those around him, in essence testing its significance and usefulness. A particularly telling example of these combined strategies occurred on 5 August 1989. Harry was in the car with the rest of the family. We were driving a few streets from home on the way to my office at the university, a route with which we were all very familiar. At one point Harry excitedly yelled out, *'Harry. Harry'*. His emphatic identification stumped us all, until we realised he was pointing at a large advertisement that dominated the skyline (p. 47). We finally recognised that he was picking out the /H/ in the advertised Hydra sign. This shows the sensitivity of his knowledge base. At that time Harry was solving the part/whole problem described above (p. 39). Similarly, on 25 September as we were travelling to the university in the car he pointed to a road sign and said, '/S/ for Sam.' He had detected the capital /S/ in the street sign, Nelson Street.

In both of these instances Harry received immediate feedback on the veracity of his detection. Our response (and each member of the family appreciated his expertise by commenting on it) was to agree there indeed was an /H/ for Harry and an /S/ for Sam. He had tested the reality of shared knowledge.

Searching for and detecting patterns: The /H/ in the Hydra sign

Harry's testing took an interesting form on occasions when he knew he was gaining control of letter formation. On 25 May 1990 Harry was writing by himself at a desk (below). He had told us he was doing 'homework', like his big brother. He generated a lot of letter-like forms and then asked us to *'Tell me what I have written'*.

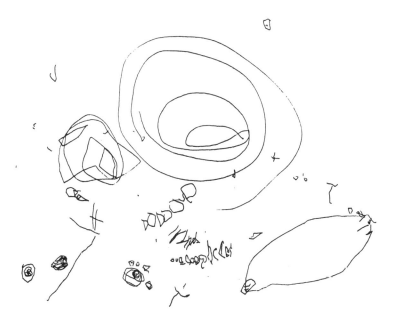

Sample 3.5 Employing strategies which generate feedback: The direction was, *'Tell me what I have written'*

An earlier example of this strategy for testing realities occurred when he said to me after he had been writing, *'Look at this'*. My interpretation of the request was that he wanted feedback on what he had written knowing that he was capable of producing elements that might have made a word. This is a very widespread strategy. Marie Clay was impressed enough with this as a common way of gaining feedback from adults that she entitled her book on writing *What Did I Write?*

Yet another explicit strategy involved analysis of the task at hand. There is a close dependent relationship between learning strategies and the developing base of knowledge in a skill. Harry knew that a task such as writing someone's name could be broken down into constituents. This meant he could search for constituent elements. The segmentation and identification of elements provided him with knowledge for this strategy, too. Ultimately it lead to what other writers have called invented spelling. The first example I recorded of this is in a sample from April 1991. He drew a seesaw and attempted to write seesaw (below).

> A close relationship exists between learning strategies and the developing knowledge in a skill. Identification of elements provides knowledge for breaking elements down into constituents.

Sample 3.6 The first recorded example of Harry's invented spelling: Seesaw

> Inventive practice involves play and experimentation with forms, e.g. invented spelling.

Inventive practice involving play and experimentation with forms was also present. It was more obvious in some phases, for example in the flourishing of constituent elements, but this too is a feature of basic learning activities. Clay (1975) identified this feature in New Zealand children's play and experimentation with the writing system at the beginning of school. She derived several developmental principles from their writing productions, including one of flexibility (experimentation with graphic symbols, creating new ones, and decorating known ones; see the samples from September and October 1989, p. 41), and of generation (using a few known symbols together with some rules for their combination to create new forms; see the sample from May and August 1990, p. 43). These principles of inventiveness have been applied to young children's writing at school entry in other countries, too (Czerniewska 1992).

Each of these observable strategies are overt forms of what Harry was doing mentally. They were focused and task-specific strategies for solving the complex problems associated with expertise in writing. They were flexible, adaptable, generative, and generalisable. They were deployed in different settings, with different people, using different materials, and with different writing content.

The Problem

But why did these learning strategies occur in the first place? What makes a child want to imitate? What triggers problem solving and pattern detection? There are two reasons.

1 Existing problems

> Children act on their environment firstly because a problem exists. The problem is to be like others, or to do as others do, to gain the knowledge and ways of doing things that are conveyed directly and incidentally by significant others.

Children act on their environment firstly because a *problem* exists. This problem (described in McNaughton 1987 in greater detail) is to be like others, or to do as others do. The 'others' are the significant people with whom the learner has interacted or those who have been observed. The problem is partially created by the child. The problem for the child is to gain the knowledge and ways of doing things that are conveyed directly and incidentally to them by significant others.

In the first volume of his autobiography Elias Canetti describes his memory of encountering a literacy problem as a preschooler. His memory is of the effect that his father's reading had on his desire to be literate. It is a memory of an encounter with an ambient activity and the consequent, fleeting joint activity.

> As soon as he began reading [the newspaper], he no longer had an eye for me, I knew he wouldn't answer anything no matter what; Mother herself wouldn't ask him anything, not even in German. I tried to find out what it was that fascinated him in the newspaper, at first I thought it was the smell; and when I was alone and nobody saw me I would climb up on the chair and greedily smell the newsprint. But then I noticed he was moving his head along the page, and I imitated him behind his back without having the page in front of me, while he held it in both hands on the table and I played on the floor behind him. Once a visitor who had entered the room called to him; he turned around and caught me performing my imaginary reading motions. He then spoke to me even before focussing on the visitor and explained that the important thing was the letters, many tiny letters, on which he knocked his fingers. Soon I would learn them myself, he said, arousing within me an unquenchable yearning for the letters. (1977, p. 26)

> Children are faced with problems which they solve in the course of acquiring expertise.

A considerable consensus exists over this notion that children are faced with problems which they solve in the course of acquiring expertise. For example, learning a language poses an extraordinarily complex problem. Just consider the dimensions of the skill that need to be picked up. The complexity of the task is daunting for theorists let alone a baby with little knowledge of linguistic textbooks.

The literacy problems facing children are multi-layered. There are the physical and structural properties of the task which constrain the forms that expert performance can take. For example, when decoding in the act of reading, or encoding in the act of writing, children need to be able to discriminate between different letters on the basis of their critical visual features and critical phonemic equivalents (Clay 1979).

But inseparable from these problems are the parts of the problem that are constituted socially by others. These include the ways in which the expertise is expressed such as how one uses letters to write one's name, the means of conveying this to children, as well as the purposes for being able to do this. What counts as a critical visual feature of a letter and phonemic equivalents is dependent on some absolute features of the visual and hearing systems. But it also depends on culturally and socially determined conventions and local expressions in family activities (see Slobin 1990, for a similar argument for language development).

I have described a general problem which one child, Harry, faced. It was that of learning to write his name. He had to learn the objective properties associated with the expertise such as how an /H/ is scripted and its critical features. But he had to learn these in the context of socially defined tasks. On some occasions, for example, the task was built around a phonemic analysis, where Harry was presented with a sequence of sound elements to map onto letters (Huh-aa-rr-ee). By presenting the writing of 'Harry' to him in this way his family created a particular sort of problem.

2 Seeing a problem

> The existence of a problem is one of the elements triggering learning.
>
> The intention to learn is a socially mediated process.

The existence of a problem is one of the elements triggering learning. But problems are problems only if someone sees them that way. The second element is perceiving that a problem exists and having an intention to solve it. At the beginnings of acquiring a skill, actions which produce learning take place because a child intends to solve what they encounter. This of course still begs the question. How is the intention to learn, for example to write one's name, created? Intention is a socially mediated process (Reed 1993; Snow 1983). What this means is the subject of the next chapter.

The Solution: In Stages?

Faced with the compelling evidence of personal agency it is tempting to describe Harry and other children as 'working out the principles on [their] own, developing a unique system, albeit one that has much in common with other children's systems' (Czerniewska 1992, p. 71). Indeed, some researchers have argued for a clear set of distinct and ordered stages which describe a single developmental sequence in early writing (Ferreiro & Teberosky 1982; Ferreiro 1985; Goodman 1990). Certainly, the physical and structural properties of the writing system and the motor and sensory systems needed for writing, together with the commonalities between children in their experience with cultural symbols may lead to some similarities.

But as Czerniewska (1992) goes on to say, 'such a picture does not seem quite adequate to explain the child's learning processes' (p. 71). This is because more variation occurs both among children in sequences and stages, and by the same child over time than fixed sequence explanations would allow.

> Co-construction explanations argue that multiple sequences follow from learning-and-development systems within different literacy activities.

The theoretical position of co-construction, which is developed further in the next chapter, argues that multiple sequences follow from learning-and-development systems within different literacy activities.[2] Within this potential for diversity is the possibility of close similarity in developmental sequences. Sometimes this similarity can be striking as shown in the comparison between two children writing their names. Harry's attempt to write Harry on 13 May 1990, when he was 2 years 6 months old, is compared with that of a child called Harriet, aged 4 years 4 months (Sample 3.7). Harriet's attempt was made in a controlled setting in New York in the early 1930s (Hildreth 1936).

Sample 3.7 Close similarities in solving a writing problem: Harriet writing in the 1930s and Harry writing in 1990

13 May 1990

> Co-construction explains similarities as similar solutions to similar 'problems' in the context of common activities.

The similarities in the representation of /H/ and the gross approximations of the letter strings almost look as though they have been made by the same child. Rather than an innate programme dictating the similarities, a co-construction framework explains this in terms of similar solutions to similar 'problems' within the context of common activities.[3]

A Theoretical Note: Views of Children

> A particular view of children, created for scientific research, sees them as active constructors of their own knowledge, but is not the only view.

A particular view of children has been employed in this description of Harry's learning. It is one which sees them as active constructors of their own knowledge. This is a view created for the needs of scientific research. It is not the only view of children possible. In Chapter 1 it was described as a legacy, in part, of Piaget's contribution to developmental psychology. However, other views have prevailed in the history of developmental psychology. For example, at times a more mechanical view has been adopted of the child as responding to the force of outside agencies, whose behaviours are shaped and constructed and who therefore has a relatively passive role in learning and development (Kessen 1979).

> Views of children's nature and their development have varied over time.

Views of children's nature and their development have varied over time. Historical analyses, for example, reveal that middle childhood as such was a 'discovery', or more appropriately an 'invention' (Kessen 1979) which took place in the middle ages in Europe. Phillipe Aries (1962) showed how families in Europe have not always identified a prolonged time of dependency and receptiveness to socialisation. But such a stage becomes apparent in the context of mass formal schooling. The general recognition and categorisation of a period of adolescence is an even more recent construction of the late nineteenth and early twentieth centuries. Among other political and social changes it was associated with the growth of apprenticeship training and formal schooling, regardless of sex and social class (Aries 1962; Cole & Cole 1989).

> Children's nature, and therefore the forms that socialisation take, also vary across cultures.

Children's nature, and therefore the forms that socialisation take, also vary across cultures. American developmentalist Howard Gardner wrote his autobiographical book *To Open Minds* as a response to the deeply challenging experiences of seeing educational practices in China. This sense of cultural difference was disturbing:

> I found myself overcome with a welter of impressions, feelings, questions, conclusions. Some of my most entrenched beliefs about education and human development had been challenged. (1991, vi.)

Much of Gardner's book is spent trying to explicate the differences between the two cultures and their views of children, learning, and development. Generalising about American child rearing he describes parents as assuming that children are born equipped to figure things out, to construct knowledge on the basis of their explorations. Parents are admonished (generally) to allow children to develop in their own way. In contrast, Gardner describes general beliefs in China as more oriented towards moulding children to perfected performance consistent with traditional modes of behaviour.

In part, a contemporary 'constructivist' rather than mechanical view has been used in this book. It is a foundation idea for the arguments in the book. Parts of the socialisation model in Chapter 1 are based on this view and some propositions about literacy development flow from it. When incorporated in a co-constructivist view both the American child-rearing beliefs and the Chinese beliefs as described by Gardner make sense.

In the previous chapter I reviewed evidence about literacy activities in New Zealand families in addition to those involved in the study I had used for the purposes of illustration (SOL, p. 19). This was done to support the generality of the propositions in the socialisation model about the role of families in selecting, arranging, and deploying activities. Unlike the previous chapter it is not necessary to establish the generality of the particular scientific view of co-construction which I have employed here to underpin the explanations. Because it is a research tool it is not an issue whether or not it matches the views of families living in New Zealand or elsewhere. The scientific view leads to an expectation that parents actively construct concepts about child rearing from multiple social and cultural sources. Different concepts would be expected in different families (see Bornstein 1991).

> Parents actively construct concepts about child rearing from many social and cultural sources.

However, because this is a book about how learning takes place, and is illustrating the theoretical arguments with descriptions from New Zealand families, it is useful to examine family views of children's learning. Moreover, an aim of the book is to contribute to more effective and equitable ways of child rearing in the culturally rich contexts of New Zealand. Therefore, it is important also to examine how the co-constructivist view matches with family views. Finding consistency or inconsistency has implications for educational endeavours built upon the socialisation model (see chapters 9 and 10).

> It is important to examine how the co-constructivist view matches with family views. Finding consistency or inconsistency has implications for educational endeavours built upon the socialisation model.

The SOL study provided some limited information. Parents were asked to describe their 4-year-old children. The request was not in terms of their theory of children, but simply, 'What sort of child is … ?'

In the context of other questions this enquiry provided insights into the parents' views of how their child learned.

The most frequent descriptors were 'active', 'outgoing', and 'independent'. Other expressions emphasised the children's inherent interest and curiosity, which bordered on these being a problem. No discernible differences existed among families in this small and selective sample.

A study by Pankhurst (1989) of preschool children's knowledge of maps also probed New Zealand parents' views of learning. A strong constructivist view was present in the parents' responses, to the extent that parents generally believed that knowledge of space and maps in particular had been actively acquired, seemingly independently of parents.

Reviews of contemporary advice to parents about children and child rearing are consistent with these constructivist views. They suggest a marked post-war (1940s) shift. An earlier view, illustrated by the beliefs of Truby King, the founder of a national neonatal care-and-advice service (Plunket), emphasised training children to be obedient and avoiding spoiling them, the goal being to establish strong parental control (Podmore & Bird 1991).

> The prevailing popular view of children has become one of children as agents of their own development, as inherently active beings who construct knowledge from resourceful environments.

The current prevailing view of children in popular manuals, on the other hand, has become one of children as agents of their own development, as inherently active beings who construct knowledge from resourceful environments (Podmore & Bird 1991). In a similar way, Piaget's 'developmental individualism' has come to underpin New Zealand schooling practice as the basis of pedagogic theory in colleges of education (Nash 1991).

These contemporary views are, however, descriptions of cultural messages in the 'foreground', which, in New Zealand, exist against a background of other messages. They are beliefs most often associated with or expressed by Pakeha middle-class families.[4]

For example, similarities and differences in emphases occur in the views of traditional and contemporary Maori educators. Metge (1984) identifies three traditional pedagogical strategies. One involved an explicit relationship between expert and selected novices in a small formal school setting and employed mainly rote learning. A second strategy was informal and embedded in the on-going life of the community as children were exposed to everyday activities. Between these was an apprenticeship strategy in which a more expert person took a selected pupil and 'fed' him or her with selected knowledge.

This diversity suggests a multifaceted view of children and their learning applicable to different educational relationships. Certainly, the traditional descriptions have a strong sense of structured socialisation and cultural channelling, but also there is a strong sense of personal construction and experimentation. This can be seen in concepts of talented novices pushing teachers (Metge 1984). It is present also in the major figure of founder hero Maui who, in several creationist episodes, actively wrests knowledge, skills, and resources from gods, nature, and parents. This mutuality of learning and teaching is expressed linguistically. Maori uses one word, ako, to mean both 'to learn' and 'to teach'. Prefixes and suffixes change the emphasis and the part of speech. But 'verbally the Maori approach stresses that unified co-operation of learner and teacher in a single enterprise' (Metge 1984, p. 2).

Contemporary educators have recast earlier models of pedagogies with contemporary voices (Irwin 1992; Nepe 1990; Pere 1991). For example, 'Te Aho Matua' (a framework of principles used for the training of Maori educators who work in Maori-immersion schools) contains elements and representations of several models.

> All forms of learning experiences provided for the child can be prompted by genuine caring and vivacity so as to prompt and stimulate in a positive way, the inner established worlds of the child's mind. ... It is an inspiration for the child to have an adult alongside as a supportive mentor throughout the learning experience. What is important, however, is that the adult must remain only as a facilitator of the child's learning, not the do-er of the task ... It is essential that children learn to sit quietly and to listen. They need to master this skill before they can hope to grasp the inner depths of the knowledge transmitted. Complementary to listening is to look, to touch, to question, to debate, to conceptualise, in order to comprehend and understand totally. (Nepe 1991, pp. 48–49)

In part, there are consistencies between the view adopted in this book and the more generally held beliefs about the active nature of children and child rearing in New Zealand families. However, the co-constructivist view affords a central role in development to social and cultural forces. This means that the view adopted here is partly at odds with generally held beliefs which concentrate on, or emphasise, the role of personal agency. It is more in keeping, though, with Maori views of teaching and learning as being parts of a whole.

Summary

In this chapter the role of the learner in co-constructing expertise has been analysed. The focus has been on the immediate mechanisms of learning and development in everyday activities. This is referred to within the more general claims of Proposition Four:

Proposition Four: Learning and development systems take form within activities *as a product of the child's actions* and the actions of significant others.

The italics in the proposition highlight the focus of the present chapter. The rest of the proposition is discussed in Chapter 4.

> Learning and development systems take form within activities as a product of the child's actions and the actions of significant others.

Implications

For families, educators, and researchers

Families: Recognising the significance of children's experiments with and play at writing is important. Their products are not mistakes and the processes are central to their development. Families can contribute to children's actions by providing them with opportunities to see family members reading and writing; by providing opportunities and materials for play; and by supporting and responding meaningfully to their writing (for example, by 'reading' their attempts, or receiving a 'letter').

> Children's experiments with and play at writing are central to their development.

Educators: Adopting a general view of the child as an active constructor of expertise carries implications for educational practices. But children's constructions are reliant on what settings provide. Providing 'rich' settings, or environments, requires careful planning to optimise opportunities for learners to observe and participate in uses, or applications, of writing; to provide guided opportunities to experiment and play with forms of writing (in which errors can be occasions for learning); and to support and respond meaningfully to children's writing.

> Children's constructions are reliant on what settings provide, so learning opportunities require carefully planned settings.

Researchers: A rich tradition exists of intensive research into children's constructions in early writing. Much of that research, however, has not systematically analysed the development of expertise in everyday activities. There is a need for systematic longitudinal descriptions of both processes and products in early writing. The framework used here also predicts that there are (at least potentially) multiple routes to similar forms of expertise.

> Longitudinal descriptions are needed of both processes and products in early writing.

Further Reading

Descriptions of early writing:

Bissex, G. (1980). *Gnys at Wrk: A Child Learns to Write and Read.* Harvard University Press, Cambridge, MA.

Clay, M. M. (1975). *What Did I Write?* Heinemann, Auckland.

Sulzby, E. & Teale, W. (1991). 'Emergent literacy'. In P. D. Pearson, M. L. Kamil & P. Mosenthal (eds.), *Handbook of Reading Research, Vol 2.* Longman, New York.

Psychological properties of expertise:

Wood, D. (1988). *How Children Think and Learn.* Basil Blackwell, London.

McNaughton, S. (1987). *Being Skilled: The Socializations of Learning to Read.* Methuen, London.

Views of children in psychology:

Kessen, W. (1979). 'The American child and other cultural inventions'. *American Psychologist.* 34, 815–820.

Maori views:

Irwin, K. (1992). 'Maori research methods and processes: an exploration and discussion'. Paper presented at the NZARE/AARE Annual Conference. Geelong, Australia.

Metge, J. (1984). *Learning and Teaching: He Tikanga Maori.* New Zealand Ministry of Education, Wellington.

End of chapter notes

1. A difficult methodological problem exists here which is formally equivalent to the principle of complementarity in atomic physics (Bohr 1958). My knowledge and intention to observe inevitably contributed to the 'everyday' context I wished to understand. This leads to the question of how the processes might have taken place without an interested observer being present. The potential for significant contextual modification is increased in the circumstances of participant observation. But it is an insoluble 'problem', however intentional and participatory the observer might be. The presence necessarily leads to a degree of 'uncertainty'. Nevertheless, I am convinced that this mode of investigation, where informed participants 'objectify' a personal developmental context is essential to understanding how learning and developmental processes work both at a microgenetic (moment by moment) and a macrogenetic (over long time periods) level. Wolf and Heath's (1992) collaborative account of early literacy within a family makes a similar point. What this 'method' requires on the one hand, is detailed reflection, but also careful qualitative and quantitative documentation to support one's claims. Where possible I have commented on how my intentions might have contributed to what I saw and what we did.

2. There are several interdependent reasons (theoretical and methodological) why a fixed and unitary sequence explanation has been adopted by researchers for emergent literacy. One has been the use of structured experimental techniques for collecting samples. These may constrain productions so that they can more easily be fitted to assumptions of fixed unitary sequences built on a theory which leads one to expect stages (Goodman 1990) or a set of core concepts (Mason & Allen 1986). Also, the prevailing theoretical position has generally led researchers to focus on knowledge and strategic performance in independent or socially restricted contexts. Asking questions about situated literacy in everyday contexts comes from other theoretical perspectives linked to a sociocultural tradition in developmental psychology (e.g. Rogoff 1990; Valsiner & Van der Veer 1993).

A final reason for focusing on the single fixed sequence is because it is derived from a normative position which sees school forms of literacy as the definitive end state or final stage of development (McNaughton 1991). Czerniewska (1992) makes similar points about research on writing, as does Valsiner (1988) more generally for research in developmental psychology.

3. This argument illustrates the complex interrelationships between the genetic potential for and social constraining of development. In this sense it is similar to two evolutionary principles used to explain similar features in animals (Gould 1989). One is a principle of homology, where similarities are due to simple inheritance and features present in common ancestors. A second is a principle of analogy. This is used to explain how the same or similar morphological features can be present in different species because of a similar response to what Gould describes as a common external push. Harry and Harriet were being reared in different historical times and culturally varied settings. The common 'push' for development existed in the similarities of their names creating a commonality in the task. It existed also in what I assume was a similarity in the socially constructed solutions to the problem.

Hildreth describes how initial or 'simple' letters emerged from her sample of children's experimentation and that the children's accomplishments were acquired, at least partly, 'in response to the parents' suggestions as to how to form letters and to spell their names' (p. 301). But Harry and Harriet, and their caregivers, are not examples of different species! Their development also reflects the commonalities of their 'cognitive endowments' and potential for learning strategies. It reflects, as well, commonalities, some genetically constrained, in how families arrange their child rearing and deploy teaching strategies.

4. A co-constructivist view has the strength of incorporating, or at least potentially incorporating, the variations in emphases that may exist across cultural groups in the messages employed in socialisation (McNaughton 1994a). It provides some means for us to understand the variations between cultures in how children and their learning are conceived, and leads us to ask questions about the sources and functions of the variations (see Chapter 9).

Chapter Four

A question of development

> **Focus**
>
> **Processes of co-construction in a writing activity**
> - The analysis of a child learning to write his name introduced in Chapter 3 is extended to include processes of co-construction.
> - This analysis shows how systems of learning and development develop within activities. These systems are a product of the child's actions and the actions of significant others.
> - Two basic and complementary types of system occur — Tutorial systems and Personal systems. These can occur in a number of ways.

How a child recognises and resolves the complex problems posed by the task of writing includes changes in knowledge and strategies.

What does Harry's sequence tell us? With the productions alone I was able to make some tentative judgements about how a child recognises and resolves the astoundingly complex problems posed by the task of writing. This includes changes in knowledge and in strategies. My ability to do this is a legacy of the pioneering work of researchers such as Ferreiro and Clay.

Ferreiro (1985) argued that this development illustrates the cognitive, or knowing, basis of the expertise. The particular sequence is a reflection of a more general sequence in literacy development which has a fixed order. She claimed that the sequence for literacy plays out in a microcosm the larger tides of cognitive development first described by Piaget (see Chapter 1, p. 12). The development in literacy parallels the movement from being relatively fixed or centred on single perspectives in thinking (preoperational thinking), to being flexible and able to adopt multiple perspectives (decentred), so that one can think about parts and wholes (operational thinking).

Ferreiro's work contributed greatly to the understanding of emergent processes. It has helped establish the general notion that development is partly a matter of gaining control over complex problems. Her work has helped delineate aspects of the problems that children might face in learning to write. It is all the more significant because she has been able to describe how the problem might look like from another's perspective — in this case the child's.

> Development is partly a matter of gaining control over complex problems.

In addition to describing the logical and formal properties of the problems that children encounter (for example, the part/whole problem) the psychological properties can be described. In a sense this means becoming even more decentred in carrying out research. What did Harry actually confront? Why did Harry want to solve the problem? Why did he engage with the activities? When this information is sought, too, it is found that writing a name involves much more than a complex symbolic problem. In an important sense Harry was taught.

> Writing a name involves more than a complex symbolic problem.

Developing Expertise: Teaching

Like many other families described in studies of children's emergent literacy (e.g. Heath 1983) Harry's family did not devise a curriculum to consistently teach him how to write his name. But it was because of our (informal) structuring of his learning that the development described in the previous chapter could take place.

One of the family influences came in the form of ambient writing activities, some of which incorporated writing names. They were salient and frequent, involving such things as writing letters to close family, writing birthday cards and notes, signing documents, and labelling materials. Harry sometimes wrote alongside other family members, or following observation of these ambient activities. The salience of these activities was underwritten by family values. Having a name and an attached symbol for that name carries important cultural meanings within the family.

> Having a name and an attached symbol for that name carries important cultural meanings within the family.

Much of the initial writing of names with Harry took place in joint activities, such as when cards and letters were signed, and when cards, letters, and other writings were read. Harry also wrote by himself, independently playing and practising writing.

The scraps of paper on which Harry wrote reveal something of his purposes in taking advantage of anything to write on: writing at banks while others were involved in transactions; writing on envelopes silently removed from desk drawers while others were writing; doing his own 'homework' on notepaper while siblings were doing theirs. Hours and hours were spent trying to write his own name and then those of other family members.

This engagement in the three kinds of activities I have described in previous chapters (ambient, joint, and personal) was socially created. In the first instance it was created by the arranging, selecting, and deploying of resources described as socialisation in Chapter 2. These resources were expressed in very direct social processes. It was through these direct processes that Harry learned how to write his name.

Two instructional interventions and their generalisation

The family members most involved in Harry's writing were his father and mother, his older brother (in August 1989, 10 years old) and sister (in August 1989, almost 8 years old), and a grandmother and grandfather who were regular caregivers several times a week. Although we could not be said to have taught him in a professional sense, nevertheless each of us functioned as his teachers. Two instructional interventions that focused on effective performance strategies show this.

One intervention came in the form of a graphemic–phonemic segmentation strategy to help Harry recognise components in his name. In a variety of ways we would say, 'Harry is spelled /H/ /a/ /r/ /r/ /y/', pointing to the separate letters as we accentuated their identification and, on some occasions, their pronunciation (Huh-aa-rr-ee). Because of what Harry said on later occasions and his generalised use of these phrases, it was apparent that Harry had internalised or appropriated the strategy.[1] He re-created the knowledge and ways of proceeding which had been created through what we had done together, our shared language, and joint focus in the activity.

In addition to this segmentation strategy I introduced Harry to a simple algorithm for producing the /H/. It took various forms, but the standard form was expressed as 'straight down, then across'. It is important to note that I had not intended to teach this algorithm. It was a response (by someone who had been trained as a primary-school teacher!) to his repeated attempts and expressed concern to control the form. As with the segmentation strategy, Harry appropriated it (rather than slavishly imitated) and employed it himself on other occasions. I recorded him saying this out loud (with no immediate audience) as he produced the sample on 18 March 1990 (p. 43).

> Strategies for breaking a word into parts and for producing a word provide a basis for generalising to other words.

Both the strategies of segmentation and the rules for production helped Harry to solve the problems that Ferreiro described. Both became part of his expertise, providing a basis for the generative strategies which were then applied to other elements and names.

On 18 August I recorded a generalisation of these instructional interventions. Harry was writing at the table while I was preparing dinner. He wanted to write something about his childcare setting, an important part of his life at that time. He said, 'How does childcare start?' I said, 'With a /C/', and wrote down a /C/. I gave him a model for the /C/, which he attempted to match.

His first attempt at imitation is shown along with my model on page 61. Harry perceived that his first attempt was not a good imitation. He found it difficult to move his pencil left in a fast arc from the first point of creating the letter. This regulatory action was expressed in his next statement. He said, 'I need to make it [the gap] fatter. I can't do it very well.' The trouble was he was starting at the bottom. Again I suggested an algorithm: 'Start at the top, then round, then stop.' His final attempts in that session are shown. The intervention with the algorithm was successful.

Sample 4.1 Generalisation of instruction: Harry writing /C/

Harry wanted to write something about his childcare. He said, *'How does childcare start?'* I said, *'With a /c/'* and I showed him.

18 August 1990

Adult model

Attempt to imitate

Harry said, *'I need to make it [the gap] fatter. I can't do it very well.'* The trouble was he was starting at the bottom.

I suggested an algorithm: *'Start at the top, then round, then stop.'*

Systems of Learning and Development

The instructional strategies which helped to construct Harry's expertise were part of a system of learning and development that had formed around Harry's growing expertise. His initial intention to solve the problem of writing had its social roots here. It was within this and in other activities that his expertise developed. In Barbara Rogoff's (1993) analysis of socialisation processes this level of analysis is described as guided participation in culturally valued activities.

The question raised in the previous chapter about why children 'see' the challenge of becoming more expert and want to become more expert can be answered. It was through guided participation. In part, children are taught to see 'the problem' and the need to be expert.

Harry wanted to write his name because we wrote, we valued it, and we spent time with him doing it. On the other hand, we were set to interpret his early and rather random marking and scribbling as strong indications of his intentions to write, in much the same way as caregivers interpret an infant's sounds and gestures as intentional acts of communication. In this way, we accorded Harry's early actions developmental significance. In this way we articulated and shaped his intention, channelling his development. We valued writing and accepted and represented his scribbles and his marks as early indicators of writing.[2]

Catherine Snow has an evocative phrase for this latent energy to act. She describes a child's intention to learn how to talk as being 'pulled out of the child' (Snow 1986). If we add the concepts of learning mechanisms, plus the guiding actions of caregivers, we have an answer to the original question of the source of intention. Intention takes shape through shared activities.

Systems evolve in joint and personal activities. Those that evolve in joint activities are called Tutorial systems. Those that evolve from personal activities I have called Personal systems. Observations by the learner of ambient activities feed into both of these systems.

> An intention to learn about reading and writing takes shape through shared activities. Systems that evolve in joint activities are called Tutorial systems. Those that evolve from personal activities are called Personal systems.

Tutorial systems

One of the first intensive descriptions of a tutorial system in early reading comes from a classic study by Ninio and Bruner (1978). They described the activity of reading picture-books (books in which a text, if present, provides single-word labels, or simple sentence frames for the label, such as *'This is a ... '*). They observed a middle-class mother in England reading with her young child from the time he was 8 months old to the time when he was 18 months.

Their observations show the mother's utterances were restricted to a few types. These occurred in a predictable order in cycles corresponding to the reading of pages. The five key utterance types were: an *attentional vocative*, which directed the child to the picture; a *query* in the form of a 'wh' question; a *label* and *feedback* about the accuracy of the child's response. The following interaction is a typical early sequence:

MOTHER	Look.	[ATTENTIONAL VOCATIVE]
CHILD	[touches picture]	
MOTHER	What are those?	[QUERY]
CHILD	[vocalises and smiles]	
MOTHER	Yes, they are rabbits.	[FEEDBACK & LABEL]
CHILD	[vocalises, smiles, and looks up at mother]	
MOTHER	[laughs] Yes, rabbit.	[FEEDBACK & LABEL]

(from Ninio & Bruner 1978)

> An activity has a structure which is shared by both participants. There are 'rules' for participation and particular routines.

Ninio and Bruner describe the predictable sequence as having a 'routinised' format. That is, the activity has a structure which is shared by both participants. There are 'rules' for participation. Within the format of the page made up of a picture and a text there is a particular routine of question, response (in the form of a label), and evaluation (or feedback).

The child's role in the activity was created within this format. It was one in which the child came to look at the picture and then labelled it, although at some stages the role could be reversed. (Later on the child could ask the mother questions, because the role and the format supporting it were so clear.) The child's role was as much a part of the format as was the mother's role. Early on she laid out the participation structure. In early sessions, after her query, she accepted and responded to any gestures and sounds of the child as instances of labelling. Her child's contribution made his role come alive. The role was learned and it developed with increasing participation.

Within this structure the child's role and his mother's role changed over time. Early on the mother accepted babbling and non-word vocalisations as if the child intended them to be words (*'Yes, they are rabbits'*). But when the child began to supply labels that were recognisable to the mother (at around 13 months) she often did not accept the previous vocalisations but demanded a correct label by repeating the query. By 18 months 50 per cent of the child's vocalisations were recognisable labels.

> The nature of the tutorial makes it a powerful pedagogical form.

This tutorial system was structured, and yet flexible. Different responses were possible within the patterns, and changes in contributions occurred over time. The presence of the structuring and the features of correction and query make this a powerful pedagogical form. Other research which I discuss in Chapter 5 has demonstrated experimentally this power for the learning of labels.

Tutorial systems such as the one Ninio and Bruner described have three closely related features. Tutorial systems are activity centred, they have a tutorial (or teaching/learning) configuration, and they are located in and express relationships (see also McNaughton 1991).

Features of tutorial systems
Activity centred

> Activities involve complex actions with particular goals.

An activity can be defined more precisely in theoretical terms. Activities involve complex actions directed towards the attainment or completion of particular goals. That means an activity is purposeful. The actions of the participants are directed to goals giving their actions intention and meaning. Among other things this means that the participants know whether they have completed the activity and how well they have done so. Participants operate strategically in the course of engaging in the activity under the conditions that are met at the time.[3]

Literacy activities may have a number of purposes for each of the participants. They have the immediate functions of the written communication, such as reading to identify the pictures in books; writing messages in a letter to relatives; and reading a piece of scripture at family devotions. In Harry's case the immediate purpose shared by Harry and his family was to enable him to write his name by himself. Other purposes included learning in particular ways and sharing the teaching/learning. More generally, our purposes were to enable him, through the activity, to become an expert family member.

> Literacy may have a number of purposes for each participant.

Purposes are not the only things expressed and constructed in the activity. Concepts about teaching/learning and about ways of acting appropriately in the family are expressed and formed through the activity. These in turn contribute to the further development of the activity because they inform and guide the participants' actions. It is at this level that links between activities can take place. One of the conditions which makes 'the problem' sensible to the child, something to overcome, is that literacy activities are embedded in family practices. Being part of the fabric of everyday life means that activities are linked with and consistent with the messages and meanings of other activities.[4]

> Being part of the fabric of everyday life means that activities are linked with and consistent with the messages and meanings of other activities.

Activities are structured in a number of ways. At a basic level this can be detected in the implicit and explicit rules for how successfully and effectively an activity has been carried out. In addition, joint activities have rules for participation which become routine ways of interacting.

Tutorially configured

For joint activities to become part of a system of learning and development the activities and the interactions have to have more than purpose and participation. They also have to have tutorial power. That is, they have to be able to produce learning. To achieve this a set of contingent, dynamic, and progressive interactions needs to be present.

> For joint activities to become part of a system of learning and development they have to be able to produce learning. A set of contingent, dynamic, and progressive interactions needs to be present which enable the child to 'see' and solve problems associated with performing the activity.

These interactions achieve important outcomes. The child's participation within the activity is made more effective or more expert than if the interactions had not occurred. Interactions enable the child to 'see' and solve problems associated with performing the activity. In so doing more effective ways of performing are constructed. Increasing competency is adjusted and responded to so that interaction patterns direct and channel the further growth of expertise. This means the child's constructive focus is continually fixed on more mature performances.

This description of the properties of tutorials is based on a developmental blueprint outlined 60 years ago which has guided subsequent research. The concept was expressed by Vygotsky in the form of a **Zone of Proximal Development** (ZPD), defined as:

> ... the distance between the actual and developmental level as determined by independent problem solving and the level of potential development as determined through problem solving under adult guidance or in collaboration with

more capable peers. .... [The concept] defines those functions that have not yet matured but are in the process of maturation, functions that will mature tomorrow, but are currently in an embryonic state. (1978, p. 86).

Vygotsky's concept provided an impetus for research into how experts and novices jointly create zones of proximal development. A report in 1976 filled in some of the detail of the working properties within a zone. David Wood and his colleagues (Wood, Bruner, & Ross 1976) described the functions of maternal behaviour when helping a child to solve construction problems. When considered as a whole the tutor's actions appeared to create a **'scaffold'**. This metaphor emphasised the supportive and constructive nature of the interactions.

> The metaphor of a 'scaffold' emphasises the supportive and constructive nature of tutorial interactions.

Since the research by Wood, Bruner, and Ross, a flurry of research activity has attempted to fill in the finer details, working on the conditions and parameters of instruction which scaffold effective performance. Several dimensions, including those noted above, have been repeatedly identified for everyday tutorials in family settings such as the picture-book reading described by Ninio and Bruner (1978).

> Tutoring provides adjustable and temporary support. The dynamic support occurs through the materials selected, the routinised format, and the changing patterns of questioning, responding, and feedback.

The tutoring provides support which is adjustable and temporary. The dynamic support takes place through the materials selected (such as picture books with static representations to aid recognition and labelling), the routinised format, and the changing patterns of questioning, responding, and feedback within the interactions. These changing patterns during picture-book reading episodes show the contingent and graduated nature of the interactions.

Within the interactions described by Ninio and Bruner one can see another major property. It is the transfer in responsibility for performance from expert to novice taking place over time. The child came to initiate labelling and even take on the role of the expert in querying about labels.

> Scaffolds enable the participants to share a view of the task that is linked to goals which both share.

Scaffolds enable the participants to share a view of the task that is linked to goals which both share. Often described in terms of joint attention, or more theoretically as 'intersubjectivity', it is illustrated in the fluidity and mutual understanding achieved within the action format. In Ninio and Bruner's study both mother and child came to know what to do within exchanges, sharing the goals of the labelling task.

Tutorials which scaffold expertise in family language and literacy are described as engaging the learner in activities that are recognisably forms of mature social and cultural uses. This means that children are engaged in something that in itself is not isolated, fragmented, and disembedded from everyday uses. In Ninio and Bruner's study this would mean that the labelling of objects, the speech event in which questions, responses and evaluations occur, the collaborative conversations about objects, and the use of books to achieve particular informative purposes are all linked, at least potentially, to other everyday ambient and joint activities. This linking or embedding may take different forms with different families. The linking of what is read in texts with other shared experiences, including those gained from other texts, is often identified with families who are white and middle class (Sulzby & Teale 1991).

Finally, the learner's performance becomes increasingly self regulated through social processes in the joint activity. In the case of the picture-book reading activity described by Ninio and Bruner it would be predicted that the child would come to know about the act of labelling, would come to know how to label in picture books, and how to check the veracity of labels, for example, *'Is that an [X]?'*

The properties of the tutorials operating in the ZPD are shown in diagrammatic form in Figure 4.1. In general, effective tutorials are associated with increasing expertise which is increasingly under the control of the learner. Over the same period of time the scaffold provided by the tutorial is dismantled and the ZPD tends to narrow and disappear as the learner is able to achieve personally what could previously only be achieved with expert guidance.

> Effective tutorials are associated with increasing expertise which is increasingly under the control of the learner.

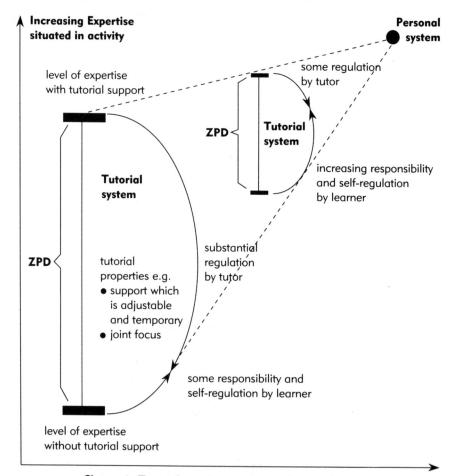

Figure 4.1 The properties of tutorials and the development of expertise within the Zone of Proximal Development

Ninio and Bruner's classic study has provided an illustration for the properties of tutorials. It is a description of formats and interactions within one type of literacy activity as developed by one dyad. Despite this it does appear to be a widespread pattern for this activity. In general, however, these properties of tutorials can be achieved in a number of ways, that is, the manner in which adjustments are made, and the ways in which performances are highlighted and support is provided can vary. This variation in the detail of the scaffold is referred to as **the specific configuration of the tutorial**.

One major way in which configurations can differ is in the explicitness of the support provided in the patterning of interactions and the language. Cazden (1993b) refers to this dimension of support as a continuum. One end of this continuum is 'telling'. The tutor can model the performance directly or tell the learner what is required of their performance.

> Tutorials can differ in the explicitness of the support provided in the patterning of interactions and the language.

This telling can occur at several levels. Specific parts of the activity can be described (e.g. *'that is the letter /H/'*) or a specific model provided (demonstrating an /H/). At more abstract levels rules can be given for how to perform. It might be possible to describe to a child quite explicitly that things can be named, that when one listens to a storybook being read one thinks about what is going to happen next and why characters do things. Whether it is desirable to 'tell' at any of these levels depends on the nature of the information, the goals of the tutor, and the expertise and needs of the learner.

It is also possible to provide explanations for performance which contribute to the support through the development of the learner's own goals and control of the activity. In Chapter 7, for example, I describe studies of an activity in which children learn to write their names. Some parents provide explanations for this activity, referring to how a name on a child's work such as a painting tells someone to whom the painting belongs. Kempton (1994) reports the following conversation during such an activity:

MOTHER Do you know what T... and S... [the child's brothers] do when they draw a picture?
CHILD   What?
MOTHER They write, when they've finished their picture, they do all their picture, when they've finished they write their name on it so they know who's done it. Sometimes the lady at crèche writes your name on your picture doesn't she? Then she hangs it up on the peg to dry.
CHILD   Yeah.

> 'Immersion' within an activity provides little direct collaboration or guidance. The problem-solving actions of the learner are heightened and much of the learning is based on personal systems evolving within the activity.

At the other end of the continuum the support is provided by 'immersion' within the activity with little collaboration or guidance. Under appropriate immersion conditions the problem-solving actions of the learner are heightened and much of the learning is based on personal systems evolving within the activity. Between these two extremes is the tutor's structuring, which enables the problem, in Cazden's terms, to be 'revealed'. The picture-book reading format described by Ninio and Bruner (p. 62) is one which enabled the general activity of labelling from static objects to be revealed to the learner, as well as enabling specific objects to be labelled.

The tutor's language can vary in another closely related way. This is the degree to which verbal elaboration of the written language that is the focus of the activity occurs. For example, in Ninio and Bruner's picture-book reading sequence, the label for rabbit could be elaborated on in various ways that go beyond the immediate association of the word with the picture. They could have talked about what sort of animal the rabbit is, rabbits the two participants had seen, or rabbits in other books. The appropriateness of doing this, at least in developmental terms, depends on such things as the expertise of the learner and the goals of the activity.[5]

Relationship-based

A third important feature of tutorial systems is the presence of a relationship. Relationships are associated with tutorials in at least three senses. These are closely related to each other. The first concerns the power or control which is expressed in the system. This dimension of the relationship between teacher and learner can be very different in different activities and even in the same activity over time. In the case of Ninio and Bruner's description of picture-book reading the relationship was one in which the mother held expert knowledge which was transmitted through the tutorial structure. The format expressed a difference in control over knowledge, although the interactional style came to be more collaborative over time as the balance shifted to the child.

The second sense in which relationships are present occurs in a set of conditions which mark effective tutorials. These include the conditions of familiarity and consistency. It is difficult to achieve the properties of scaffolded tutorials without coming to know a lot about the other participant(s). Being able to construct a dynamic scaffold is dependent on intimate knowledge of the child's performance within the activity over time. This means that familiarity and consistency of engaging in the joint activity are important determinants of how effective the scaffold can be. This is one reason why effectively scaffolded tutorials are harder to achieve in classrooms.

The third dimension to relationships is the presence of a positive affective base. This positive emotionality is partly a platform on which the tutorial system can be held together. A close emotional bond provides a basis for sensitive contingent responding. But it is partly an outcome of the effectiveness of the initial tutorial itself. Indeed, Bronfenbrenner (1979) has argued that children come to enjoy being with older family members (they develop attachments with these members) because of their engagement in joint activities.

> Familiarity is an important determinant of how effective a scaffold can be.

> A close emotional bond provides a basis for sensitive contingent responding.

Types of tutorial systems

Patterns of interacting within tutorial systems can differ in a number of ways. Some patterns produce qualitatively different types of tutorials. Differences can occur in how frequently the patterns occur, too. These different patterns define the configuration of the tutorial. Different configurations typically are associated with different sorts of literacy activity. But differences can occur within the same activity over time at different phases in the development of expertise.

Three very general types of tutorial configurations are described here. These types have very different formats and interaction patterns. They construct learning and development in very different ways. In each of them the relationship between the tutor and the learner is different. But variations in how these features operate within the tutorial can determine the effectiveness of the tutorial and the patterns of learning and development within the activity.

> Different configurations can occur within the same activity over time at different phases in the development of expertise.

1. Collaborative participation

In this tutorial configuration of collaborative participation the child interacts with the more expert person. They perform the task at hand together. The child's performance is often described as one which has been negotiated and this reflects the 'give and take' quality of the relationship. The focus of this negotiation is often the meanings in what has been, or is being, written.

> In collaborative participation the child's interactions with the more expert person have a 'give and take' quality. The expert's control reduces over time.

The manner of interacting and the way in which the task is presented are initially controlled by the expert, but that control reduces over time. This means that the relationship takes place and develops within a supportive framework. A typical example can be found in studies of reading storybooks to preschoolers which is described in more detail in the next chapter (the example comes from Phillips & McNaughton 1990).

A mother and her 4-year-old read a new storybook (*What Made Tiddalik Laugh?*, Troughton 1977) over several weeks. The narrative structure of this Australian story has a central or precipitating problem. A giant frog drinks all the animals' water. The result is that the animals have a problem which needs to be solved — how to get the water back.

The frog's act occurs close to the beginning of the book. When we tape-recorded the joint reading we found that the mother and her child paused at this point. Initially, it was the mother who interrupted reading and highlighted, through the ensuing conversation, the problem precipitated by the frog. This is shown in the transcript of the first reading reproduced below. In later readings, when the book was familiar, the child interrupted at the same point in the text. By the time the second exchange occurred, which is shown below (about three weeks later), the mother had reduced her direction to her daughter and prompted her to identify the central problem (*'What do you think ... ?'*). Both exchanges have the properties of collaboration and negotiation of the text's meanings. They show supportive conversational and textual framework and the shifting balance of control over the negotiation.

TEXT: *In the dream time there lived a giant frog called Tiddalik. One morning, when he awoke, he said to himself, 'I am so thirsty, I could drink a lake!' And that is what he did.*

First reading:

MOTHER Goodness! He's a big frog.
CHILD And look at all those little frogs.
MOTHER Yes [laughing]. They're all looking at him. Probably ... going, 'HUUUUU. What an enormous one!'

Later reading:

CHILD I wonder what those frogs are looking at him for?
MOTHER Why do you think?
CHILD I don't know.
MOTHER What! Do you think they're a bit surprised at how ... large he is and what he's doing?
CHILD I think ... well ... he's eating all their water.
MOTHER Yes, I do too.

The mode of the interaction is often described in educational writing as an enquiry mode, although this places emphasis on the child's actions. The learner is not told directly what to do. So, in terms of Cazden's (1993b) analysis of the explicitness of support the scaffold 'reveals' what is to be acquired and the learning is embedded in the collaborative framework. The scaffold involves indirect mechanisms for cuing, highlighting, and channelling performance.

This means that the focus is on making sure that the expert's understanding of the task becomes the learner's through the negotiation of meanings. The focus is on gaining expert knowledge of the way the performance or the skill works. The focus is also on the learner coming to control the performance. Children are intimately involved in this performance as conversational partners and as decision makers, having rights of participation and initiation that are similar to those of the expert.

Two developmental effects of this way of interacting are important. The first is the development of verbally mediated control and reflective, or thoughtful, performance. This means that the dialogue style of participation enables a growing verbal control over the performance of the skill. For example, with storybooks, children come to reflect on what is being read and check their understandings. The second related outcome is the potential for generalisation of the strategies in this verbally controlled performance to similar situations or performances.[6] Children apply their strategies for reflection and checking when listening to new storybooks.

Other developmental outcomes are examined more fully in Chapter 5. These are the social and cultural values which develop from such interactive experiences. Among these are a view of the informative world as containing resources that can be exploited; and being socialised into a child-centred view of teaching and learning.

2. Directed performance

The second major tutorial configuration, of directed performance, utilises and expresses a very different relationship. In this configuration children are involved as performers within the supporting conditions created by the expert and the activity. The expert's role is to model the performance. The learner's role is to imitate the model, to match the performance. Explicit collaboration and negotiation between participants of the performance does not typically take place.

The manner of my teaching Harry how to form an /H/, and the algorithm for /C/ are good examples. My role was to present a clear easy-to-follow demonstration. I modified the demonstration, exaggerating and attenuating my demonstrations contingent on his production. I might even have ended up physically guiding him through the performance had his motor skills not been controllable through my language. So even though explicit collaboration may not take place, the scaffold as a whole is socially created, both directly and indirectly, as my teaching of Harry shows.

The expert's construction comes from the choice of materials, the flexible presentations of chunks of the model, and the systematic modification of models and feedback contingent on the accuracy and detail of the performance. The learner's contribution is in the reconstruction of the performance, constructive

> In a directed performance the expert's role is to model the performance. The learner's role is to imitate the model, to match the performance.

analysis of patterns, and reflections on the match. Harry's contributions then involved guidance of his own attempts, examining the visual features and motor components involved in making an /H/, and checking and modifying his attempts against models. Just as with collaborative participation, the tutorial provides a structure within which the performance can occur.

In literacy activities one effect of these tutorials is to produce accurate performance which matches the tutor's model. The focus is not on enquiring about the meanings in a text.[7] It is on achieving a good rendition. Similarly, the focus is not on increasing the potential for generalised performance to other situations. Indeed, that would be a problem. The letter /C/ is a /C/ in whichever context it appears and has to be rendered in a particular way (give or take the vagaries of English orthography). So the focus is on getting it right. In many respects the relationship and the learning is built on a clear authority, the authority of the demonstrator. If the previous (collaborative) configuration (p. 69) can be described as an apprenticeship to understanding meanings, this one can be described as an apprenticeship to accurate reproduction of a model.

Other developmental effects of this directed performance tutorial are important to note. They include the development of learning skills associated with imitating. In the context of the general conditions of tutorial systems, which include the embedding of the activity in family practices, it can be predicted that effective tutorials enable the learner to become an expert imitator. The learning skills that develop include being able to represent and remember in ways that facilitate matching the performance.

3. Item conveyancing

The third general configuration is concerned with accumulating items of information. The learner is involved as recipient or demonstrator of that knowledge. In one sense this is an indirect version of the preceding configuration of performance direction. However, rather than a performance being the focus of the tutorial, it is the acquisition of a piece of information or a way of acting which is required to be publicly displayed.

The dynamic properties of the tutorial are to be found in the patterning and modification of questions and feedback. Again, task and materials form part of the supporting structure as do the contingent properties of the interactions.

The manner of conveying information entails an authoritative relationship, because the expert holds authority by virtue of having knowledge. A bit of information is to be learned and the tutorial is focused on acquiring and displaying this knowledge. What may happen within the tutorial structure, as I noted earlier, is that the balance of power in the relationship may shift.

Ninio and Bruner's study (p. 62) provides a clear example of this configuration. The general pattern of the interactions can be described in terms of a Query–Response–Feedback sequence, with a fourth term, the initial direction to orient to the page (attentional vocative) as an optional early component. Writers have pointed out the similarity of this form to many classroom interactions in which the teacher asks a question and evaluates a student's response (Heath 1983; Wells 1986). Reviewing research on classroom discourse led Cazden (1988) to identify this as an I–R–E sequence (**Initiation–Response–Evaluation**) which seems to be the 'default' or fall-back condition of classroom lessons.

What may be different in the case of this configuration in families compared with that in classrooms is the changing role of the learner (from novice to holding authoritative knowledge within the family). This is associated with greater changes in the expression of the relationship between family members and the growing child.

Within this item-conveyancing configuration children learn the specific items of information. They also learn to display those items. Ninio and Bruner argued from their study that the child learned labels. But also he learned that the world can be partitioned up and labelled, and he learned strategies for finding and checking labels. For example, he could appropriate the I–R–E sequence either directly or indirectly in exchanges within other activities. In this sense the tutorial configuration can lead to some generalised expertise, for example, knowing that one can find labels for things.

> During item conveyancing the role of the learner changes from novice to holder of authoritative knowledge.

Tutorial configurations: a summary

What I have described summarises, in an idealised form, three types of configurations for tutorials. Teaching (being an agency of socialisation) is an occupation in which one tries to put oneself out of business. It is also difficult because the degree of support for students and the dynamics of the interactions need to vary over time. This is to be expected if an effective learning and development system is functioning. However, supportive interactions may be more or less effectively managed at any particular point in the developing skill. For example, I could have misread entirely Harry's motor co-ordination and frustrated him terribly by continuing to tell him how to do a /C/. As the exchange unfolded it appears I didn't, which is attributable to my intimate familiarity with our ways of learning together.

These three tutorial configurations are not mutually exclusive or mutually antagonistic ways of teaching/learning. In later chapters I describe how some families may emphasise one configuration over another in emergent literacy. But other families show remarkable tutorial dexterity employing multiple forms at different times to suit different purposes in literacy activities. Expert and learner can shift into and out of different configurations within and across activities and tasks. These variations are described in chapters 5, 6, and 7.

> The three tutorial configurations are not mutually exclusive or mutually antagonistic ways of teaching/learning.

Three qualifications need to be made in this analysis of tutorials. The first qualification is that components of tutorials are fluid and configurations can shift dramatically within and across activities. Categorising different configurations may freeze this fluidity and create an inappropriately rigid view of configurations being incompatible. What has been argued here is that these configuration types create qualitatively different ways of teaching/learning, and these ways need to be evaluated in terms of criteria for effectiveness which are associated with the goals of the activity as co-constructed by children and experts.

The second qualification is sometimes not recognised as writers praise particular forms. Different systems may be well or poorly co-constructed. Tutorials of any configuration may vary from being effective to ineffective.[8]

> These qualitatively different ways of teaching/learning in tutorial configurations need to be evaluated for effectiveness in terms of the goals of the activity.

A QUESTION OF DEVELOPMENT

The third qualification is related to the first two qualifications. It concerns the role of extended language within each of the configurations. In each of these types some elaboration on and negotiation about the performance or focus can occur. Further interactions at these points add to the developmental characteristics of the tutorial.

The participation properties of the three tutorial type are shown in Figure 4.2

Figure 4.2 Three configurations of tutorials showing changes in participation over time by the Tutor (T) and the child Learner (L), leading to two uses of the tutorial by the Learner.

Time 3	Personal (Inner) dialogue	Learner uses routine as strategy with tutors	Personal (Inner) dialogue	Learner uses routine as strategy with tutors	Personal (Inner) dialogue	Learner uses routine as strategy with tutors
	L↔L	L→T	L↔L	L→T	L↔L	L→T
Time 2	T↘L T↗L Shared responsibility in conversations		T→L←T Learner taking over more of the performance		T→L←T Learner responding more accurately or completely	
Time 1	T↗L T→L Initiation-Response (Conversational Routine) Initially tutor takes more responsibility		T→L→T Model-Demonstration-Evaluation (Performance Routine) Initially tutor takes more responsibility		T→L←T Initiation-Response-Evaluation (Display Routine) Initially tutor takes more responsibility	
	Collaborative Participation		**Directed Performance**		**Item Conveyancing**	

> In collaborative participation each action is dependent on the last. Over time the child becomes more able to take responsibility for clarifying and extending or negotiating meanings.

In **collaborative participation** the interactions look like conversations, with each action dependent (contingent) on the last. Over time the child becomes more able to take responsibility for clarifying and extending or negotiating meanings. A point is reached where the child is able to do what could only be scaffolded jointly. So, for example, in storybook reading, which is focused on gaining narrative meanings, a point is reached both within familiar books and even with unfamiliar books where the child can engage in an inner 'dialogue' asking about and elaborating on meanings. Also there comes a time when the child can check meanings strategically with others, for example, by explicitly asking about meanings (shown as the second usage at the top of Figure 4.2).

Directed performance tutorials have a three-part structure in which a model is presented (**Model**), performed (**Demonstration**), and checked (**Evaluation**). Over time the child's performance becomes fuller and the degree to which the model is needed reduces. Again, the shift to relative independence is expressed in two ways: the self-monitoring and checking of performance by the child, and the child's development of imitation.

The classic three-part I–R–E sequence (p. 72) is the basis for **item conveyancing.** Over time the child's response becomes surer and the need for corrective feedback reduces. Again, the development from the basic structure to relatively independent functioning can be seen in two forms. Firstly, the child comes to initiate his or her own queries about information and check items in a personal inner dialogue. Secondly, others can be used strategically to gain the needed items.

Personal Systems

A second major type of learning and development system forms around literacy activities. It takes place as children engage in literacy activity by themselves. The term independent is avoided here because in one sense it is not independent at all. Just as with joint activities these activities have close links to other activities. They are carried out relatively autonomously, but their existence and some of their features are derived from ambient and joint activities.

> Personal systems develop when the learner performs their growing expertise without the guidance of others.

Personal systems develop when the learner performs their growing expertise without the guidance of others. The child mechanisms entailed in these systems include forms of imitation.[9] Children 'collect' the things they have seen, and experience in joint activities, and experiment and play with them. The forms this experimentation takes are based on what children have observed and experienced.

Other mechanisms include what Vygotsky described as self-teaching, that is, the learner creates the properties of a tutorial system based on what they have constructed in the joint activity. This means personal systems are self-configured, rather than configured by a joint tutorial structure. In this sense the child as learner can be said to create their own Zone of Proximal Development (see pp. 64 & 65), their own 'scaffold' (Wertsch 1985).

> Personal systems are self-configured; the learner creates their own 'scaffold'.

Apart from the presence of a 'significant other', the properties of Personal systems are similar to those of Tutorial systems. They are activity-centred and learner-actioned. What is different about these systems is that performances and learning are inherently more variable than those taking place in joint activities. The self-generated configurations are much more open and flexible than those occurring during joint activities. In joint activities the variation is more limited by the rules of participation and other structures that have been socially created (see pp. 64–68). In play it is the child who determines the structure.

This accounts for some of the stages in the development of Harry's expertise. For example, his flourishing of forms was an outcome of systematic trial and error and persistent imitation. These processes were functionally linked to the intention to become expert and the presence of both significant ambient activities and powerful joint activities.

Harry's independent practice with his name was activity-centred. He wrote deliberately to achieve the correct writing of his name both for its own sake and to complete activities associated with writing one's name. He wrote to add his name on birthday cards. He wrote to add his name on letters. His experimentation with writing his name was structured by the activity, for example, by writing on lines. And it was structured by the model (for the /H/ and for his whole name) remembered from interactions within the Tutorial system (pp. 59–62).

The development of expertise in Personal systems can be very impressive. Until recently, much of the research on emergent literacy has been focused on this impressiveness, how children appear to build knowledge and expertise from ambient activities. Several intensive case studies, such as the one reported in Glena Bissex's (1980) book *Gnys at Wrk*, document personal constructions, in Bissex's case in the form of her child's invented spelling.

> Peer groups can provide a significant context for both Tutorial and Personal systems to develop. Peers also can amplify conflict or the sensing of problems.

Peer groups can provide a significant context for both Tutorial and Personal systems to develop. Vygotsky claimed that more capable peers can create a Zone of Proximal Development. But it is also the case that peers can amplify conflict or the sensing of problems. In doing both of these things, Personal systems can be activated. The following scenario, from a set of observations of an informal peer group in a kindergarten, shows fleeting examples of Tutorial systems with peers and of personal construction through peer feedback (from Gray 1994).

A group of 4-year-olds are playing with clay. The teacher is present.

FERGUS   [makes a huge /F/ and an /e/]
FERGUS   [leans over to look at Hugh's letter /h/] Lookit, lookit, Hugh made his h.
HUGH   You can make this into an o. [points to Fergus's letter /e/]. I'll show you. [leans over to show him]
FERGUS   (does not respond to Hugh, but continues to make, remake, and redefine his /o/ shape). Teacher ... teacher, I can't make my real name ... [pause] ... teacher?
TEACHER [looks but makes no comment; a large /F/, small /e/ and an /o/ are clear]
HUGH   [makes tiny indentation in his dough, mumbles to no one in particular] What can I make now?
FERGUS   Look! [said to Hugh] I made mine into bridges of the bridge. Now it's all lines. See I changed it! See? Children look! [loud shrill voice; his letters have been turned into cylinder shapes and laid across each other]
TEACHER [smiles] What have you made Fergus?
FERGUS   These are ... these ... see ... they are bridges of the bridge. [points, voice high and excited]
TEACHER [smiles, touches the clay shape] Children, Fergus has made a good representation of bridge struts, which he calls bridges of the bridge.

In this example Fergus played with making letter shapes. A critical point came when his experimentation did not produce a letter which satisfied his model for making his 'real name'. He changed his letters into line shapes and redefined their identity and function. But in the process he observed that letters can be broken or modified into constituents which include lines.

The potential for a Personal system arose at least at two points in the process of making letters — when he tried to overcome a perturbation in his performance, and when he reformulated his letter shapes. His own scaffolding to produce his name was not sufficient to meet his own requirements for performance.

Summary

This chapter has explored the co-construction processes which occurred when Harry learned to write his name. It has used this analysis to explore two further propositions:

Proposition Four: Learning and development systems take form within activities as a product of the child's actions and the actions of significant others.

Proposition Five: Two basic and complementary types of system occur and each can be expressed in a number of ways.

I have used this analysis of the co-construction process to identify the general learning–teaching mechanisms at work in the socialisation of emergent literacy. The next chapters apply this analysis to specific reading and writing activities.

> Learning and development systems take form within activities as a product of the child's actions and the actions of significant others.

Implications

For families, educators, and researchers

Families: What families do in providing ambient activities, in engaging in joint activities, and in creating opportunities for personal activities makes a difference to the development of expertise.

Educators: Adopting the view that development is co-constructed carries significant implications in addition to those noted in the previous chapter. Among these are a concern to monitor and evaluate how educational settings provide experiences of ambient activities and joint activities, as well as opportunities for personal activities which meet effectively the setting's educational goals. A major implication of this view is that any assessment of an individual child's development is, at least indirectly, an assessment of the actions of both educators and of children. Individual profiles of children's expertise are needed, but so too are profiles of joint actions within tutorials.

Researchers: Among the challenges facing researchers is providing comprehensive analyses of co-construction within systems of learning and development. This requires intensive moment by moment (microgenetic) analysis, as well as analyses of patterns of change in the system over periods of time, as well as demonstrating and evaluating effectiveness of systems. In addition, there is a need to show how processes connect across activities and how children internalise (or appropriate) expertise from tutorials.

> Providing ambient activities, engaging in joint activities, and creating opportunities for personal activities promote the development of expertise.

Further Reading

Descriptions of tutorial properties:

Cazden, C. (1988). *Classroom Discourse.* Heinemann, Portsmouth, NH.

Cazden, C. (1993b) 'Immersing, revealing and telling: a continuum from implicit to explicit teaching'. Paper presented to the Second International Conference on Teacher Education in Second Language Teaching. City Polytechnic of Hong Kong.

McNaughton, S. (1991). 'The faces of instruction: models of how children learn from tutors'. In J. Morss & T. Linzey (eds.). *Growing Up: The Politics of Human Learning*. Longman, Auckland.

Wood, D., Bruner, J. & Ross, G. (1976). 'The role of tutoring in problem solving'. *Journal of Child Psychology and Psychiatry*, 17, 89–100.

Processes of internalisation (appropriation):

Rogoff, B. (1993). 'Children's guided participation and participatory appropriation in sociocultural activity'. In R. H. Wozniak and K. W. Fischer (eds.). *Development in Context: Acting and Thinking in Specific Environments*. Lawrence Erlbaum, Hillsdale, NJ.

End of chapter notes

1. Rogoff (1993) uses the term 'appropriation' to refer to the process by which individuals actively transform their expertise through participation in activities. By making the shared process their own they can engage in future situations in new ways. The learning processes described earlier as cognitive endowments contribute to this process. The term appropriation extends or elaborates on the process of internalisation described by Vygotsky (1978). Problems do exist, however, with the use of either term, which both Rogoff and Valsiner (1994a) have discussed. The major problem concerns the ways in which the concept of internalisation has come to denote a simple dualism of outside forces/inside constructions which writers such as Rogoff and Valsiner wish to avoid. Rogoff argues for appropriation stressing notions of participation and mutually constitutive processes. Yet Valsiner argues for keeping a sense of separation between personal and social, referring to a concept of 'inclusive separation'. My use of both terms (internalisation and appropriation) is to emphasise concepts of participation and the role of the learner.

2. This is a view of intention as socially constructed. Reed (1993) states the position clearly: 'For social creatures like human beings the formation of intentions is frequently an interpsychic and dialectical process in which more than one individual is involved, and in which social constraints and cultural meanings play a major role.' (p. 46). This position carries important educational implications such as the question of how broad or narrow the 'funnel' of a caregiver's or teacher's perception of intentional acts might be. Parents typically have wide funnels compared with adults who are not familiar with, and not responsible as caregivers for, the infant.

3. Contemporary analyses of activities draw on earlier theoretical writings on activities from Leont'ev and others (see Wertsch 1985; 1991). Three levels for analysis are identified: general motivation for the activity as a whole, specific goal-directed actions, and the operational characteristics of action under particular sets of circumstances. Rogoff (1993) describes the analysis of activities as:

> ... a shift away from considering cognition as a collection of mental possessions (such as thoughts, schemas, memories, scripts and plans) to regarding cognition as the active process of solving mental and other problems (e.g. by thinking, recounting, remembering, organising, planning and contemplating) generally in the service of intelligent action. ... Mental processes such as remembering, planning, contemplating, calculating or narrating a story generally occur in the service of accomplishing something, and cannot be dissected away from the goal to be accomplished and

the practical and interpersonal (as well as intrapersonal) actions used.... An important feature [of this approach] is its embedding of individual activity in specific sociocultural contexts. (pp. 124–125)

4. This point is related to note 6 at the end of Chapter Two. The essential heterogeneity of messages in socialisation is paralleled by multiple expressions of the same meanings in different activities. This means that there is high redundancy in messages available to children about, for example, how to do things and why they are done that way (Valsiner 1994a; Wertsch 1991).

5. The language between participants in a tutorial has been analysed in additional ways which show how differences in their usage contribute to the developmental properties of the tutorial. One is the identification of elaborations as 'non-immediate' or 'decontextualised' language (e.g. explanations of storybook elements and/or their significance). Snow and her colleagues (e.g. Snow 1991) have examined the significance of the use of this sort of language for later progress at school in literacy activities focused on comprehension. The degree of negotiation between participants has been another focus for analysis. Negotiation between participants occurs, for example, as meanings of stories are discussed or as agreement over how to help or what sort of help is needed occurs. The negotiation contributes to the learner's reflecting on the activity and to social and cultural messages about the nature of written language (see chapters 6 & 7).

6. The question of how to promote generalisation has occupied researchers for decades. One of the major determinants of transfer is the degree to which the learner acquires the means to reflect on and articulate features of their expertise and effective conditions for their learning. Enquiry-type exchanges in collaborative participation provide a social base for reflection with the potential outcome of enhanced generalisation (see Bransford 1979; and Rogoff 1990 for discussion and analyses of enquiry-type exchanges, which demonstrate both of these potential outcomes for verbally mediated collaborative tutorials).

7. In many of these examples of performance tutorials in literacy activities described in this book, there is little verbalised negotiation or elaborative or explanatory talk. However, the actions of the participants in such tutorials may include these verbal extensions. The examples of writing activities described in Chapter 7 show several instances of explanatory and elaborative talk around the meanings of the activity, even given the performance focus. Similarly, often tutorials which occurred or occur in cultures that do not use a written language are assumed to be exclusively focused on performance. Clearly, this is not the case, and explanatory talk in oral narratives and even in the formal learning of important tribal memories may include explanations and asides (see Feldman 1991; Metge 1984).

8. Van der Veer and Valsiner (in press) point out that many writers have enthusiastically assumed that Vygotsky's blueprint for socially mediated learning both represents the best way to learn, and also is generally effective. The authors note that this promise of an 'educational utopia' seen in Vygotsky's writings has created a 'blind spot' in the analysis and application of his work to tutorial systems. These authors are not alone in commenting on the uncritical acceptance of sociocultural models of tutorials (e.g. Damon 1991; McNaughton 1994b).

9. The form of imitation which Baldwin called 'persistent imitation' is a fundamental process in both joint activities and in personal systems. Features of the model are changed through the modification of its representation in a dynamic process of circular reaction.

This means that the child acts on what has been modelled (either directly in joint activity or indirectly in ambient activity), in acting modifies the model, gets feedback about the imitation from personal or socially mediated checking, modifies the emergent representation, which provides a platform for further action. Harry's attempts at /H/ were modified forms of our models, which provided opportunities for him or us to check similarities, which facilitated further attempts. This process enables the development of innovative forms in the present through continuity with previous activity (see Valsiner 1994a; Vasconcello & Valsiner 1993).

Part Three

Early Activity Systems

Chapter Five

Early activity systems: reading

> **Focus**
>
> **Activity systems for reading**
> - The descriptions in this chapter are based on four of the propositions introduced in the first chapter: Propositions Three, Four, Five, and Six.
> - A framework for describing activity systems for reading is presented.
> - The framework is based on descriptions of the participants' ideas, the interaction patterns, the tutorial types, and the materials used in the activity.
> - A set of activity systems for reading is described, together with the typical expertise that develops within the systems.

This chapter describes activities from which reading skills develop. My intention is similar to the intentions underlying other systematic scientific descriptions, such as Ferreiro's (1985) descriptions of types of solutions to the problems of learning to write. It is to discipline the information which is available from case studies, controlled descriptions, and experimental analyses. This chapter tidies these loosely connected bits into a framework to provide educators and researchers with a tool for developing programmes and research. The framework is shown in Table 5.1 (p. 83).

The framework is based on the identification of different activities and their associated learning and development systems. It identifies: the ideas (including goals) about the activity which the participants have; the patterns of interactions that take place; and the types of tutorial typically involved in the activity (the tutorial configuration). The typical materials used are identified also because they

> The framework in this chapter identifies the ideas about the activity, the patterns of interactions, and the types of tutorial typically involved.

contribute to the structure of the activity. Using the socialisation model that has been built up in previous chapters, predictions about typical consequences associated with specific activities can be made. This means that activity systems are identified which, under appropriate circumstances, enable certain forms of expertise to develop. The four dimensions that provide the framework are described first, and then specific activity systems are identified.

Four Dimensions of an Activity System for Reading (TABLE 5.1)

1 Ideas & Goals

> Goals may be relatively explicit and clearly formed or relatively general and not easily explained to someone else.

The first dimension of an activity system for reading concerns the goals that the participants have for the activity. Family members and children may intend to achieve a variety of goals, including the solving of particular literacy 'problems', such as identifying the letter /H/. These intentions may be very explicit and well formed. In reading a picture-book of animals many readers deliberately aim to have their child learn the names of animals. And they can describe this aim to an

Table 5.1 A framework of four dimensions for describing an activity system for reading. (The example comes from page 98)

Ideas Goals	Interaction Patterns	Tutorial Type	Materials
Negotiating narrative meanings and pleasure and knowing conventions	Conversational/narrative routine within dyadic exchanges and display routine within dyadic exchanges	Collaborative participation and (embedded) item conveyancing	Storybook

MOTHER	The cat went on until he met some land crabs. The land crabs said, 'Get out of the way cat or we will clip you'. Look how fat the cat is.
ROSIE	Fat up here.
MOTHER	Ummm. Turn the page. Where do I start reading? Look at the page. Which side ... Which side does Mummy start reading?
ROSIE	That side.
MOTHER	It's this side here. Always this side [taps page] ... Okay? Show Mummy where, where I start reading.
ROSIE	Here.
MOTHER	Right ... right from the top. Then go down, and then onto the side afterwards. The cat said, 'I've eaten a hundred cakes, I've eaten ... '

interested observer. But intentions can also be relatively unformed or general and not easily explained to someone else.

There is a complication in identifying intentions. It arises because intentions may not be shared. At the beginnings of an activity a function of the tutor is to develop shared goals (Wood, Bruner, & Ross 1976; Wertsch 1991). This means that one criterion for an effective tutorial is the degree to which jointly shared goals have been established. A number of researchers have described how teachers and pupils in beginning a reading activity may not have developed a shared view of the nature of the tasks in which they engage. The consequence is relatively ineffective instruction and learning.[1]

> At the beginning of an activity a function of the tutor is to develop shared goals, otherwise instruction and learning are relatively ineffective.

A further complication is that the same activity may serve different purposes. This is illustrated in the purposes for which storybook reading has been used in our family. Among other things, reading stories has been used to occupy children while waiting for a bus; to ease the transition from our care to a babysitter's care; to relax a child before sleep; and to provide a distraction from a household mishap. These are in addition to the goals associated with getting meanings from a book.

Goals may be identified through self-report by learners as well as family members. They can be inferred also from features of the exchanges that take place between learners and others engaged in the activity. One feature is the degree to which exchanges are routine and take place without confusion. A second related feature is the clarity of the negotiation that takes place. A third feature is the degree to which the learner can adopt the role originally taken by other family members. For example, can a 3-year-old used to reading picture-books for labels ask questions (such as *'What's that?'* [in the picture]) of other participants, including younger family members?

It needs to be repeated here that goals are important expressions of identity, crystallising messages about who one is and what a child needs to become. But goals are not the only significant ideas that caregivers have. There are also ideas about how to socialise the child and the forms that teaching should take. These are, in turn, related to more global ideas about the nature of learning, development, and the role of teaching (see Chapter 3, pp. 53–54).

2 Interaction patterns

> Interactions in tutorials provide predictable frameworks for learners and experts to influence each other, and become familiar ways of behaving within the tutorial.

One of the properties of interactions in tutorials is that they provide predictable frameworks for learners and experts to influence each other. Interactions take place that become familiar ways of behaving within the tutorial. These familiar exchanges have been called 'participatory routines' by Peters and Boag (1987). Their concern has been to analyse how language interactions enable children to learn the purposes and meanings of their culture's language. They define a routine as ' ... a sequence of exchanges in which one speaker's utterance, accompanied by appropriate non-verbal behaviour, calls forth one of a limited set of responses by one or more of the other participants. ... [They] ... specify the content and kinds of utterances to be expected' (p. 81). Routines also contain 'participation structures' which specify who can say what to whom.

The interactions in Table 5.1 are described as involving narrative routines and display routines within dyadic exchanges. Rosie and her mother (the dyad) participated in familiar exchanges. One, focused on elaborating meanings, entailed a conversation about the illustration related to the storyline. The other focused on conventions and employed the I–R–E sequence.

The important property of these routines is the provision of a format within which interactions become familiar and repeatable. This creates a structure for what is said and done as well as how it is said and done. Such routines are effective to the degree that they have become habitual yet provide a flexible platform for changes in the child's responsibility and expertise.

> Effective routines are habitual, yet provide a flexible platform for learning.

I have introduced examples of routines in literacy activities already. Among the most commonly reported are those routines first systematically described by Ninio & Bruner (1978). Their descriptions of a mother reading picture books with her child have all the hallmarks of a routinised performance (see Chapter 4, p. 63). Particular sorts of routines typically are associated with particular goals and are the building blocks for particular sorts of tutorials.

How routines take place (including who participates and how), and their content, are aspects that carry social and cultural messages. In Chapter 2 (pp. 31–33) I described how different participation structures which occur when reading storybooks to children carry cultural meanings.

> How routines take place, and their content, are aspects that carry social and cultural messages.

3 Tutorial types

The third dimension of an activity system for reading is the patterning of the learning and development system that develops within an activity. In Chapter 4 (p. 68) the different patterns were referred to as tutorial configurations. Three major configurations were described: **collaborative participation, directed performance,** and **item conveyancing.**

The core properties of tutorials need to be kept in mind when describing an activity system — they provide support which is adjustable, temporary, and dynamic. Responsibility for the performance gradually transfers from expert to novice. Participants come to share goals. The tutorial activity is recognisably a form of mature social and cultural uses. Finally, the learner's performance becomes increasingly self-regulated.

> Tutorials provide support which is adjustable, temporary, and dynamic. Responsibility for the performance gradually transfers from expert to novice. Participants come to share goals, and the learner's performance becomes increasingly self-regulated.

Additional tutorial properties discussed in Chapter 4 (p. 67) included the manner in which language is used to support learning and to negotiate participation in the activity. Each of these features carry important cultural messages about how one learns, as well as what one learns as a member of the family.

4 Materials

The final dimension used in the classification of a reading activity system focuses on the materials employed in the activity: texts, such as storybooks, picture books, and message books (such as books with religious lessons); signs, such as labels on cereal packets, menus and traffic signals; and visual display items, such as letters on a chart and flash cards. A fourth category can be identified in the oral-language activities that have literate structures or impinge on literate knowledge. These include rhymes, word games, and narratives. It is beyond the scope of this book to deal with these in depth but it is important to recognise that these, too, are activities within which systems of learning can develop that bear on literacy (see Goswami & Bryant 1990; Snow 1991).

> Rhymes, word games, and narratives are activities within which systems of learning can develop that bear on literacy.

Exploring Activity Systems for Reading

The rest of this chapter describes activities for reading. I apply the four dimensions above to Tutorial systems, and then discuss their relationships with Personal systems and the development of specific expertise. The descriptions show how the four dimensions contribute to the overall nature of the activity. If these activities are examined moment by moment in families it is found that family environments and classrooms share a common feature — activities can shift rapidly over time.

Although some activities are relatively fixed from beginning to end with little change in how they are structured, sometimes rapid shifts within activities can happen. On some occasions the majority of the time is spent in a single activity as defined by the materials, the types of tutorials, the interaction patterns, and the ideas of the participants. On other occasions embedded activities occur which can change the overall nature of the activity, as it were in mid flight. A shift in an activity is signalled by a change in goals, in routines, and in the type of the tutorial.

> A shift within an activity is signalled by a change in goals, in routines, and in the type of the tutorial.

Table 5.1 (p. 83) summarises these four dimensions with an example from storybook reading. An episode of storybook reading could be described in terms of the ideas of the participants (to construct narrative meanings but sometimes identifying conventions, e.g. concepts about print); the typical interaction patterns (mostly the conversational turn-taking of narrative routines by mother and child with some I–R–E sequences — see Figure 4.2, p. 73); the tutorial type (a mixture of collaborative participation with embedded item conveyancing); and the material (storybook).

Reading for labels

Tutorial systems

Labelling activity is clearest during picture-book reading. Almost without exception researchers have used similar descriptions of reading picture-books to very young children. The dominant intention of caregivers has been inferred from the routine and the structure of the tutorial. It has seldom come from stated intentions because researchers typically have not asked what these were.[2] The goal is to teach children particular labels and to teach that labelling of the static objects in the text is possible (the development of reference, or lexicalisation). Less obviously, as children become more familiar with picture-book reading the goal is to provide elaborated descriptions of objects, adding to their meanings.

> Labelling activity is clearest during picture-book reading.

The standard routine, illustrated in the classic study by Ninio and Bruner (1978), is a labelling or, more generally, a display routine. Its typical characteristics were described in the previous chapter. In its earliest form the routine has a formal structure similar to the I–R–E sequence. Given that both tutor and child are attending to the same picture then a /wh/ question usually occupies the initial position (*'What's that?'*). This is followed by a label (*'An [X]'*); and then feedback (*'Yes, an [X]'*) or elaboration.

This sequence is highly predictable. But over time the complete sequence with its supporting structure shifts. At first, the reader initiates and even responds for the child. Later on, child-initiated sequences develop.

The example of labelling routines in Sample dialogue 5.1 comes from a Samoan family in the SOL study described in earlier chapters. A picture-book of animals was being read with the 4-year-old Jerry. The reader was his school-age brother Robin, but the boys' father was present as was the baby, Ben. Sections from a long session of reading are shown to illustrate modified and mixed forms of the basic I–R–E sequence together with elaborated exchanges which can develop from that base.

> The reading-for-labels sequence is highly predictable, but over time the complete sequence with its supporting structure shifts. At first the reader initiates and even responds for the child. Later on, child-initiated sequences develop.

Sample dialogue 5.1
Modified labelling routines

Reading an animal picture-book

JERRY	Crab	(LABEL)
ROBIN	No, that's a squid, that's a squid, that's a crayfish. That's a centipede.	(FEEDBACK)
	Oh, look at that, Jerry. That's a centipede.	(ATTENTIONAL VOCATIVE & LABEL)
JERRY	Oh, what's that?	(QUERY)
ROBIN	A garden spider.	(LABEL)
	What one do you like?	(ELABORATION & EXTENDED CONSERVATION)
JERRY	I think that one there because they have a big thing.	
ROBIN	I like this one. You know why? …	
JERRY	Cause this one's bigger than any kind of thing. They grow more bigger than you. They grow more bigger than any kind. … This is a crawling fish. A crawling fish. We can buy some of that.	
FATHER	Crayfish.	
JERRY	But … up here a crayfish … this is a crayfish.	

The example shows the 4-year-old taking over initiating exchanges which earlier may have been initiated by his older brother. He initiated these with a query (*'What's that?'*) and by labelling (*'Crab'*). It also shows the use of conversations which elaborate or extend the meanings of the labels. This extended conversation takes on some of the features of collaborative participation. The two boys were building meanings through the conversation. The participation structure was dyadic within a multiparty framework. At the point of some confusion about a label the boys' father provided an authoritative statement. This reflects the multiplicity of caregiving roles and status which are a more general characteristic of some Samoan oral language exchanges (Ochs 1982).

The type of tutoring typically associated with picture-books is **item conveyancing**. Early in development and early in the reading of a particular book the conveyancing takes place in a controlled and directed fashion. There is an obvious reason for this. Knowledge of specific labels and the knowledge that things can have labels come from social sources.

In the absence of relevant knowledge there is nothing in a picture to tell you what a thing might be called. That information has to come from someone who 'knows' the thing. The teaching/learning functions are carried in the prompts ('*What's that?*'), models ('*It's an [X]*') and specific feedback, including corrections ('*No, it's not an [X]*').

Ninio and Bruner's (1978) study (see p. 63) did not demonstrate experimentally these developmental outcomes through controlled manipulation. But their descriptions show the child's expert labelling was associated with graduated changes in the scaffold provided by the mother. For example, when the child was able to say some words the mother did not accept a gesture or a non-word vocalisation as a good approximation for a label. She would prompt for a more recognisable label ('*What's that?*').

A number of descriptive and correlational studies support this conclusion (Sulzby & Teale 1991). In addition, a programme of research by Whitehurst and his colleagues has provided experimental support. In their original study (Whitehurst *et al.* 1988) a group of mothers from middle-class families was instructed to increase two sorts of interactions. They were to increase their use of progressive labelling routines and to increase extended conversations which elaborated on the labels in a way similar to that of Jerry and his brother. Compared with similar mothers and their 2- to 3-year-olds, these mothers achieved both goals. They both extended their conversation around labels, encouraging children to talk about the pictures, and provided more informative feedback in the labelling routines.

The results showed immediate effects in the children's increased ability to label common objects after four weeks of reading. But also it showed more generalised effects of the elaborative exchanges. These children had longer and more expressive language in story conversations which continued to be more extensive than the other children's language over the ensuing six months.[3]

> Picture-book reading may develop into more sophisticated interactions involving information in books. Whether this happens depends on further ideas and selections of activities by family members and the child.

Picture-book reading may be the precursor of more sophisticated interactions focused on information in books, depending on further ideas and selections of activities by family members and the child. For example, children's knowledge of labels and meanings can be checked and elaborated on by others when reading storybooks and more complex expository texts. In these texts the routine is focused on meanings and definitions carried in the text rather than pictures that depict objects. The focus is to check the child's understanding of the label (e.g. '*Do you know what a volcano is?/ What is a volcano?*'). Definitions, discussions, and linking with personal experience may follow. Questions about this form of the activity have only recently begun to be asked, and it is apparent that at least some children beginning school are aware of how one reads expository texts to gain and extend meanings (Pappas 1993).

> The picture-book is a particularly compelling sort of text.

No other major tutorial types or routines or intentions have been described with picture-books. The picture-book medium is a particularly compelling sort of text. Readers are very likely to adopt the naming goals, basic interaction patterns, and tutorial configuration described here. This has been shown in research across different countries and with different sociocultural groups (Sulzby & Teale 1991). The patterns are very robust and occur despite differences in how familiar the readers and children are with the topic and even how typical the format of the book is (Pellegrini *et al.* 1990). Where variations are identified families have been described as producing fewer or less elaborate labelling exchanges than is typical in white middle-class dyads, and fewer progressive adjustments to elicit labels. This is most obvious when the comparisons have involved families from lower socioeconomic groups (e.g. Ninio 1980).[4]

Personal systems and the development of expertise

The initial activity for children of reading picture-books involves at least two participants. But the social strategies for labelling (I–R–E) and elaborating on meanings in conversations of collaborative participation can become personal. In becoming personal they provide the intention and means to search for and construct meanings in other texts. Although illustrations may not occur in the book to provide a context for assigning labels, texts do provide routes to word meanings through linguistic structures, including grammatical rules as well as the relationships between words.

From the original social basis, children can assign meanings, and elaborate on the meanings of words (their connotations), given textual support and the personalised strategies for assigning meanings, even without elaborated discussion. Precursors of this connotative defining can be found in the initial routines of simple labelling (which provided denotative definitions). This outcome may have particular significance for the transition to school.

Often called incidental learning from text, this elaboration of word meanings is particularly significant for acquiring new vocabulary during the school years. Warwick Elley (1987) has shown how primary-school children listening to stories being read by their teachers improved their knowledge of unfamiliar words. A particularly significant vehicle for this occurred when teachers discussed new words in passing. But even children who heard the stories without embedded discussion significantly improved their understandings of some words, although not as much as the group who experienced embedded discussions.[5]

Reading for narrative meanings

Tutorial systems

> Storybooks are the typical medium for reading for narrative meanings — they are texts which have a narrative structure.

Storybooks are the typical medium for reading for narrative meanings. Storybooks are those texts which have a narrative structure. A typical narrative structure contains an initial setting, an explicitly stated problem, a series of actions performed by the characters in a series of episodes, and a resolution of the problem. This is not an exhaustive and exclusive definition. Rather, it describes standard features that distinguish stories from other sorts of texts, for example, picture-books. In the case of picture-books, a definition highlights how these texts contain list-like collections of pictures and labels with little connected text and with no explicit theme.[6]

Because of the assumed significance of storybook reading to further literacy development I have devoted the next chapter to discussing variations in activities and outcomes, particularly as they express social and cultural meanings. In that chapter I examine how many ways there are to read storybooks. Two major activities, with some variations on each, have been described in the research literature. One of these is the activity being focused on in this section, reading for narrative. The second activity is identified below as reading for performance.

The major purposes reported by caregivers for reading for narrative include pleasure and enjoyment, for parents as well as children. Parents also report a goal for children to develop knowledge about the world and about books. These general goals come from both middle-class families (e.g. Phillips 1986; Wolf & Heath 1992) and other sociocultural groups. Some of the families in the SOL study, however, added the more focused comment that, for them, reading to their children was associated with preparing them to do well at school.

When interaction patterns between readers and children are analysed the routinised content and forms suggest another set of purposes, which are closely tied to ideas of what it means to understand a story. These interaction patterns strongly suggest that readers and children come to share an intention to build up meanings and comprehend the narrative through discussion. Included in this set of purposes are: identifying the goals and central problems of a story, clarifying the meanings of text, integrating and anticipating meanings, and linking texts to personal experiences (see Chapter 6).

Storybooks are not the only texts that can be read with conversational (or more specifically, narrative) routines. A narrative may be formed around picture-books. For example, a consistent connected narrative can be added to some of Richard Scarry's texts, which have little text but do have thematically linked illustrations.

The typical interactional patterns between readers and children within this activity involve narrative routines. Although less formalised, more open ended and often longer than the basic patterns associated with labelling, nevertheless they have a routinised structure. They begin with comments, questions, directives, and statements inserted into the text and require responses from partners that are semantically contingent on (develop the meanings from) the preceding item of interaction.

The sample dialogue on page 91 comes from a mother and 4-year-old child the first time they read a book entitled *Ahhh Said the Stork* (Rose 1977). In this narrative the precipitating problem is the presence of an unfamiliar egg.

The mother's initial insertion of a query into the text prompted the child to anticipate possible events in the story. But the child's hesitation called forth an adjustment in the query from the mother in the form of a more direct prompt of general knowledge (*'What comes out of eggs?'*). It is important to note that the adjustment relied on the child's level of understanding; it provided more direct information as well as following the meaning. The exchange continued with the mother agreeing with the child's response (*'Sometimes'*) and following this she provided feedback and elaboration. A feature of these extended conversations is the links made between the text and the child's knowledge, including knowledge of other texts.

The habitual features of these narrative routine exchanges are that they focus on some aspect of meaning which is shared and negotiated. The 'rules' are hard

> The typical interactional patterns in storybook reading begin with comments, questions, directives and statements inserted into the text and require responses from partners that develop meanings.

Sample dialogue 5.3
Narrative routines

Reading a storybook

TEXT *While they were thinking the egg made a croaking sound. It wobbled and tiny cracks appeared.*

MOTHER	What do you think might come out of there?	(QUERY)
CHILD	Don't know.	(RESPONSE)
MOTHER	What comes out of eggs?	(QUERY)
CHILD	Birdies.	(RESPONSE)
MOTHER	Sometimes, but there are other things that come out of eggs too. Don't they? Remember the platypus? Maybe we'll see what comes out of an egg. [turns page] Hullo, hullo hullo.	(ELABORATION)

(from Phillips & McNaughton 1990)

to write because of their open-ended nature. But in general they involve a question or comment which provides an orienting focus, followed by a semantically contingent response, and further responses leading to agreed or shared meaning. Given the agenda to develop appropriate meanings which the expert reader already may know, then these sequences sometimes come close to being an I–R–E sequence (see Figure 4.2, p. 73). But whereas the item-conveyancing routine functions primarily to display knowledge for checking, the sequence in Sample Dialogue 5.3 functions to elaborate on meanings in the text, making them shared, and able to be checked and refined.

> Narrative routines are like conversations in which the rules for participation are dynamic and flexible.

The participation rules are dynamic and flexible. Early in storybook reading it may be that the reader initiates the majority of exchanges, but with experience the 'right' to insert a comment or question into the text comes to be shared. The supporting structure that this exchange provides is apparent in the illustration from *Ahhh Said the Stork*. In the conversation the mother draws on her familiarity with her child's knowledge, essentially their shared knowledge (*'What comes out of eggs?'*). These interactions enable effective adjustments to be achieved, ultimately providing slots for the mother to provide or anticipate appropriate meanings. In the example above she ends the exchange not by supplying an answer but by raising the child's anticipation and dramatically highlighting the answer as *contained in the text*. The total exchange illustrates those scaffolding functions of pinpointing critical features in performance, of demonstrating strategies (*'Maybe we'll see ... '*), and other functions of tutor's behaviour such as maintaining a consistent focus (Wood, Bruner, & Ross 1976).

> Changes take place within and between books as the novice acquires more of the expertise which is nascent in the narrative routine.

Further analysis of these exchanges occurs in the next chapter. They share similar developmental properties with others I have discussed. Changes take place within and between books as the novice acquires more of the expertise which is nascent in the routine. They are the building blocks for the tutorial which takes the form of collaborative participation. As reader and child collaborate in discussing meanings the structure shifts with increasing responsibility for identifying meanings devolving to the child. The significant variations in the use of language for support and negotiation of meanings are discussed in the next chapter. Over time the amount of conversation may reduce as the child appropriates the shared dialogue and uses it themselves.

Personal systems and the development of expertise

What develops from this storybook-reading activity? Evidence for several developmental consequences can be found (Sulzby & Teale 1991; Teale 1984). One is the development of what has been called a story schema or story grammar. This involves identifying and using the typical structure of written stories, including features of written language such as the register of storybooks, which is different from conversational speech. Another is the comprehension of narrative. Children learn to make connections and develop meanings with a particular text. But, generalisable strategies for comprehension also develop. These include the search for and clarification of text meanings and themes, as well as the linking of text meanings with experiences, including those from other books. An example of this occurred in the exchange between the mother and child reading *Ahhh Said the Stork* described earlier ('... *Remember the platypus?* ...'). Yet another outcome is concepts about the conventions of print and how books 'work'. Such concepts typically do not arise directly from the general activity, but from embedded activities usually under the control of the learner (see 'Reading for items and concepts', p. 97).

> Storybook reading has several developmental consequences: the development of a story schema or grammar, comprehension of narrative, generalisable strategies for comprehension, and concepts about conventions of print and how books 'work'.

Reading for performance

Tutorial systems

Storybooks can be read to children in a manner entirely different from that described above, resulting in a very different activity. Children can experience texts in a way that results in them performing the text. In this activity children come to recite passages that have previously been read to them. I describe in Chapter 6 how families read storybooks for performance and I explore what it means to read storybooks this way. It is a relatively uncommon documented feature in the published international research, apparently reflecting the cultural and social groups with whom researchers have worked. In New Zealand it appears to be relatively common with some groups.

The more common texts with which performances take place are religious texts (particularly hymns, Sunday-school stories and Bible passages) and nursery rhymes. An example of the former from a group of Tongan children learning a hymn at Sunday school is described in Chapter 10 (p. 183). The Samoan family conducting their daily devotions (lotu) described in Chapter 2 read Bible passages in this way with the young members of the family (p. 25).

A partial-performance routine is shown in the sample opposite of a mother and 4-year-old reading from a nursery rhyme-book. The example comes from a Pakeha family in the SOL study (Chapter 2). The segment comes part way through the full rhyme (*I Know an Old Lady*) and shows the mother providing a gap at the end of lines. She marked this gap as an invitation for the child to complete (performing the text) by rising intonation and a pause. There were several problems in the child's performance and several instances where the mother repaired the incomplete or inaccurate performance. The presence of the repairs, and the fluidity of the demonstration and imitation (such as *'Bird ... Bird'*) underscore the well-rehearsed nature of the exchange. The 4-year-old knew what to do, the mother knew what to do, and they cued each other.

Sample dialogue 5.4
Performance routines

Reading a nursery rhyme (segment)

MOTHER	*I know an old lady who swallowed a ... ?*	(PARTIAL MODEL)
CHILD	Hog.	(PERFORMANCE)
MOTHER	Dog.	(REPAIR & MODEL)
	It's a dog, that's not a hog.	(& FEEDBACK)
	What a hog to swallow a dog.	(MODEL)
	Yes, you knew that didn't you?	(FEEDBACK)
	She swallowed the dog to catch the ... ?	(PARTIAL MODEL)
	Cat.	(REPAIR & MODEL)
CHILD	Cat.	(IMITATION)
MOTHER	*She swallowed the cat to catch the ... ?*	(PARTIAL MODEL)
CHILD	Spider.	(PERFORMANCE)
MOTHER	Bird.	(REPAIR & MODEL)
CHILD	Bird.	(IMITATION)

The intention of the more expert readers in these examples (including the Sunday-school teachers in the sample dialogue in Chapter 10, p. 183) is to produce an accurate rendition of what has been read or demonstrated. Samoan families who teach children texts in this manner report a concern to have children learn texts literally *'by heart'*. The goal of establishing the text as an authority is discussed further in Chapter 6.

At the heart of the activity is a performance routine. The typical sequence can be seen when first learning a text. It has three components: a Model followed by an Imitation, and explicit or implicit Feedback about the accuracy of the imitation or performance. In the nursery-rhyme sample dialogue above (5.4) this Feedback occurred through the repairs. The chunks of text that are modelled are increased over time. This allows longer and longer sections of text to be performed between successive imitations, which may become more like shorthand prompts. At an intermediate stage much of the text is able to be performed without models or reduced prompts. The more expert reader repairs hesitations or mistakes using a truncated version of the sequence and a correct word or phrase is modelled for imitation. The third stage of the learning is a fully accurate performance by the learner.

As learning takes place the role of the expert includes responding to the growth of independent performance by extending the amount of text modelled. The typical shifts in expertise involve the novice in becoming able to recite or perform the text by themselves. This development occurs within the structure provided by the expert demonstrator, and through the joint focus, where the significance of the task is shared. In its ideal form, then, this tutorial configuration of **directed performance** has all the hallmarks of scaffolded instruction. It provides dynamic, flexible, and temporary support. The functions

> The typical performance-routine sequence has three components: a Model followed by an Imitation, and explicit or implicit Feedback about the accuracy of the imitation or performance.

of the tutor are to finely tune and adjust the models to support and extend performances.

Personal systems and the development of expertise

Just as with the labelling activity (see p. 89), two levels of expertise are developing: the level of learning the particular text, and the level of learning how to learn via directed performance — learning that it is possible to learn passages through such a configuration. In this way, novices learn how to imitate, they learn that expert performance can be imitated, and that one can be guided through those performances.

An example of the adaptations and development of this tutorial configuration comes from a Samoan family in the SOL study. We visited Sa at home when she was 5 years 6 months old. Her mother was writing out a number of Bible verses (in Samoan) for a Tala (a religious story or play at Sunday school). Then she attached these to pieces of paper for the children, including Sa, to learn by heart.

The immediate psychological consequence for literacy development here includes the growth of what can be called recitation memory. This is strategic memory for texts, and also for memory within the directed performance type of tutorial. Extensive and generalised experience with performance tutorials with texts is likely to have consequences for the development of recitation memory. Limited support for this prediction appears in Wagner and Spratt's (1987) study of Moroccan children. They compared cognitive outcomes for children in Quranic preschools compared with several groups of other Moroccan children. The pedagogy in the Quranic preschools consisted of large amounts of recitation and memorisation of verses from the Quran in Arabic. Compared with children in other preschools not emphasising this pedagogy, and children not going to preschools, the Quranic preschool children had superior performances on specific cognitive tasks to do with serial memory (such as memorising lists of names or digits). The patterning of results suggests that the specificity of the profiles (increased serial memory but not on other cognitive tasks) was due to the recitation pedagogy, not Arabic literacy as such.

More generally, these recitation and memorisation routines and tutorial configurations are reported as occurring in the language interactions that take place between parents and children learning to talk in some cultures. Given that the children come to be expert language users, these contribute to the effective socialisation of language development (Schieffelin & Ochs 1986). However, there may be some important variations in the use of the tutorial in different tasks. For example, recitation was, and is, a significant element in preferred Maori pedagogy (Metge 1984). Metge points out, however, that recitation in formal learning of waiata (songs) and whakapapa (genealogies) was, and is, a means to an end. Adaptation and innovation to suit the needs of particular audiences are also features of that expertise. And in the course of formal demonstrations discussion about meanings may be interposed between segments of recited text to enable novices to develop the means for taking into consideration the variables of time, place, and audience (see Chapter 6).

Testimony to its adaptability comes from early accounts of literacy learning

> Extensive and generalised experience with performance tutorials with texts may have consequences for the development of recitation memory, but not necessarily for other cognitive tasks.

> Recitation and memorisation routines contribute to the effective socialisation of language development in some cultures and are an element in preferred Maori pedagogy.

by Maori following early contact with church missionaries and colonisers. The power of learning via demonstration is shown in its deployment in a new context. The account here is from the English missionary W. R. Wade in 1838.

> The rapidity with which the alphabet, catechisms, and elementary lessons have been learned by men and women who before had no conception of a book, is familiar to those who have visited the mission stations in New Zealand, or who have read published accounts of such visits. A day or two after our landing on Missionary grounds in the Bay of Islands I was introduced to a chief, who sat beside me with a portion of the New Testament in his hand. He appeared to be reading, verbatim seriatim, following correctly verse after verse. To my great surprise, I learned that he had no knowledge whatever of reading; but having a clue to the commencement of the chapter, his memory accomplished the rest. ... A woman advanced in years, even to grey hairs, and blind in one eye, and who was only just aroused to enquiry on the subject of religion, came to converse with me. I put a plain question to her; when, instead of giving me a straightforward reply, she repeated, for an answer, a whole chapter out of the New Testament.

Although Wade excludes these demonstrations as reading, other accounts by Maori do recount how directed performance tutorials and performance routines provided a vehicle to learn to decode print (see illustration on p. 96).

Reading for messages

Tutorial systems

The activity of reading for messages occurs most obviously with materials called 'message' texts. Religious texts are archetypal message books, but, in general, message books are those texts that have been written to convey a truth, an important moral, a lesson for living, or those texts that are read as though they contain such authoritative lessons. Aesop's fables, Grimms' fairy-tales, and Sunday-school or church stories contain messages which can be read as moral imperatives. The dominant purpose in the activity is to establish the nature and authority of the message. It is not only to promote the learner's comprehension of that message, but also to establish acceptance and perhaps a commitment to a moral or lesson.

The primary purpose of establishing the authority of the message is expressed in an agreement routine or in a version of a performance routine (discussed earlier, p. 92). In the agreement routine the person reading seeks agreement or acceptance from the listener. An example, from one of the Samoan families in the SOL study, is provided in Sample dialogue 5.5.

> Message books are texts that have been written to convey a truth, or that are read as though they contain such authoritative lessons. The dominant purpose is not only to promote the learner's comprehension of that message, but also to establish acceptance and perhaps a commitment to a moral or lesson.

Sample dialogue 5.5
An agreement routine

Reading from a church book
FATHER   When I go to Sabbath school, I oh so quiet be. I will never push or run. Oh just you see.
         Is Mummy happy when I obey? Oh yes Mummy is happy when I obey. She smiles and says, 'You're a good boy today.' Is Dad happy when I obey? (MESSAGE)
CHILD   Yeah. (AGREEMENT)
FATHER   Oh yes, Daddy is happy when I obey.

An account by a Maori of learning to read in the early colonial period, and an illustration of the activity by a contemporary colonial artist.

(*The Maori and His First Printed Books*, A. H. & A. W. Reed 1935. Courtesy Hocken Library)

MAORIS LEARNING TO READ

"We learnt every day, every night. We sat at night in the hut, all round the hut in the middle.... Sometimes we went to sleep upon the book, then woke up and read again. After we had been there six months, we could read a little, very slowly."

A minimal type of tutorial is created from this simple interaction. Nevertheless, forms of expertise develop within the bare structure. The tutorial can be described as a variation of the performance-directed type. In reading the text 'Sabbath School' the father reading the text asks his child to agree with the message. Heath (1983) describes white working-class families reading similar texts and also minimally interacting. To the extent that the message is explicit, or is assumed to be explicit by the participants, then no further interaction takes place. Over time the child's accurate memory for and comprehension of the lesson of the text develops.

Personal systems and the development of expertise

What forms of expertise develop within the activity of reading for messages? There is little in the way of research evidence to guide an answer. A possible consequence, however, would be memory for particular lessons, and, more generally, the development of generalisable strategies for comprehending lessons. These strategies would enable the listening child to identify the lesson, the central moral, in essence, strategically answering the question 'What should I remember?'

Consequences of reading for messages may be memory for particular lessons and the development of generalisable strategies for comprehending lessons.

Reading for items and concepts

Tutorial systems

> The activity of reading for items takes place when alphabetic information is specifically identified in texts.

> The major purpose of reading for items is to teach recognition and identification of letters and words.

The activity of reading for items takes place when alphabetic information is specifically identified in texts. Typical materials are alphabet books and books which have single words or captions on pages, as well as charts and letters affixed to walls and other household surfaces. This activity can be embedded in texts used with any of the previous reading activities.

The major purpose of the activity is to teach recognition (discrimination) and identification of letters and words. Parents articulate this as an aim. They can nominate a general developmental expectation (see Chapter 2) and deliberately act in accordance with this aim. For example, in the SOL study we found that parents expected their children to be learning to identify the letters of the alphabet around the age of 4 years 6 months. Those who had this expectation deployed resources to create Zones of Promoted Action. They had books, and in the case of the Samoan families, wall charts. On a test of letter identification (Clay 1979), the children, on average, could identify seven letters at 4 years 6 months, and 15 letters at 5 years. The Samoan children on this test could identify nine letters and 24 letters respectively. When we used the wall chart they could identify many more letters. Those parents who said they did not expect their children to begin to identify letters until they got to school were not trying to teach them and the children had lower identification scores on the test.

The routine associated with this activity is a labelling routine (p. 86). They occur within the tutorial configuration I have called item conveyancing (Chapter 4, p. 71). As with labelling, the routine has the characteristic three-part sequence (*'Can you see the /H/ for Harry?'* [child points] *'Yes, there's the /H/.'*). This early structure later gives way to a child-initiated identification (*'There's an /H/ for Harry'*). In its ideal form support is provided by the predictability of texts in terms of the items and simple language, and of the routine.

> Participation in the reading of books provides opportunities for children to build concepts about print, and about structural properties of books.

Concepts about written language can be acquired in similar ways. Participating in the reading of any type of book provides opportunities for children to build concepts about print. The term 'concepts about print' refers to knowledge children come to have about structural properties of books. These include knowing that a book is orientated in a particular way, that print rather than illustrations contain the message, and that print has direction, e.g. left to right (see Clay 1979).

The activity sometimes takes a more explicit form in reading storybooks. An example from the SOL study is shown on page 98. A Samoan mother and her daughter aged 4 years 6 months were recorded reading the storybook *No More Cakes*. A segment from the middle of the reading is shown. A typical narrative routine which maintains the focus on the narrative theme (the cat's voracious appetite) is followed by routines focused on concepts about print. Explicit labelling routines were used (*'Where ... ?'* – *'Here'* — *'Right'*).

This is an unusual example. These questions had never appeared before in the transcripts. They appeared shortly after a session in which the child's concepts about print had been tested. It was our research practice not to exclude the family when we were probing this knowledge. In this particular instance the mother watched our questioning and then incorporated it into her own practice. In a sense we created an ambient event from which the mother constructed messages about appropriate pedagogical aims for book reading.

Sample dialogue 5.6
Labelling routines for concepts about print

Reading a storybook (segment)
MOTHER *The cat went on until he met some land crabs. The land crabs said, 'Get out of the way cat or we will clip you.'*
Look how fat the cat is.
ROSIE Fat up here.
MOTHER Ummm. Turn the page. Where do I start reading? Look at the page. Which side ... Which side does Mummy start reading?
ROSIE That side.
MOTHER It's this side here. Always this side [taps page] ... Okay?
Show Mummy where, where I start reading.
ROSIE Here.
MOTHER Right ... right from the top. Then go down, and then onto the side afterwards.
The cat said, 'I've eaten a hundred cakes, I've eaten ...

Personal systems and the development of expertise

More typically, children acquire knowledge of concepts without explicit conveyancing and no formal tutorial is associated with this activity. It is a construction by the child embedded in other activities and revealed as the learner actively builds knowledge and solves 'problems'.[7]

Similar developmental outcomes occur in reading for items as they do for labelling (p. 86). However, in addition to knowing that things can be labelled (in another context of use), the item knowledge provides children with strategies to link graphemes (letters) with phonemes (sounds). It is a key for unlocking other sorts of knowledge. A repeated finding in studies that have followed children from home to school is that knowledge of letters predicts early progress in learning to read (McNaughton 1987; Tizard *et al.* 1988).

Other activities

Texts designed for children provide immediate, purpose-built materials for tutorials. But other materials are available in the immediate environment. When I first read Dolores Durkin's (1966) observations of children who learned to read before school I was impressed with her own reflections on just how rich the ambient environment is for children's learning about literacy. She became aware of the extent to which children are immersed in written language. Among the more salient materials within which children are immersed are signs. Signs provide information about events, and things to do. They stand for or signal, rather than describe. The obvious examples are labels of foodstuffs, traffic signs, and advertisements.

Activities involving signs can take a direct form in which someone intends to

> Signs are among the more salient literacy materials within which children are immersed.

identify letters or words. They can also occur under the initiation of the child as part of the developing personal system for exploring ambient information. In both instances the activity has most of the characteristics of the activity of reading for items. The differences are in the materials and the structuring of the activity. In the case of reading signs the activity structure is provided by the familiarity of the event within which the sign occurs.

The signs which provide the materials for this activity occur as part of the familiar events within the family. Having breakfast and using cereal in a packet, or driving to a childcare centre under the control of traffic signs have other purposes. However, they have a sequence and a predictable habitual event structure. Breakfast may not appear orderly, but in disorderliness there is predictability. And those participating can choose to notice the item information in the ambient environment.

I have provided several examples of activities using signs in earlier chapters. They include Harry's discussion with his grandfather about reading Marmite (see Chapter 1, p. 13), his identification of the /H/ in a Hydra bacon sign (see Chapter 3, p. 46) and children in the SOL study commenting on words at the breakfast table (see Chapter 2, p. 28).

These examples show the labelling routine in its developed form under the learner's control. The tutorial has become driven by the child's intention within the structuring provided by the familiar sign and event. In both of the instances with Harry he initiated the exchange that took place. The Hydra example shows him adopting part of the labelling routine. Doing this in diverse situations was his forte at this time. He searched and found /H/ for Harry all over the place. In this instance he located and used the attentional vocative 'look', but rather than ask the /wh/ question he simply asserted, 'There's an /H/ for Harry'. We all adopted an audience or expert role agreeing with him and applauding the assertion (once we had identified what he was referring to).

Other forms of identifying items can occur too. Alternative forms can be distinguished from reading items in texts on the basis of the materials. Among the materials which families use to teach items are flash cards and alphabet charts. The same routines and tutorial configurations I have discussed in this chapter are applicable here. Expertise within these very structured activities tends to be even more closely bound to the activity structure than in some others.[8]

Summary

In this chapter I have described several activity systems for reading. The goal has been to examine systematically the structures and developmental significance of these activities. Four propositions summarise what has been described:

Proposition Three: Literacy practices are expressed in specific **reading** activities which have identifiable constituents. These include goals, rules for participation, and ways of carrying out the activity.

Proposition Four: Systems for learning and development take form within

reading activities as a product of the child's actions and the actions of significant others.

Proposition Five: Two basic and complementary types of system occur — the tutorial and the personal — and both can be expressed in a number of ways.

Proposition Six: What children learn to do with written language is become relative experts within particular **reading** activities.

In Chapter 7, I examine activity systems for writing guided by the same four propositions. Before I do this, however, I need to examine that very special reading activity — reading storybooks to children.

Implications

For families, educators, and researchers

Families: The activities that families promote are important sites for learning — for both children and the family 'experts'. Family members have important roles in children's learning and development in reading, and their ideas, beliefs, and goals are important also. Although this requires 'resources' these do not have to be material resources, nor expensive. The everyday and seemingly ordinary things one can do with children (for example, reading a book together, family devotions) are the important occasions. Beginning to do a range of different things with reading is an important goal (for example, understanding and talking about stories, playing with rhymes, memorising passages, and reading signs).

> Family members have important roles in children's learning and development in reading. Everyday and seemingly ordinary events are important occasions for learning.

Educators: The ideas and goals families have for their children's expertise and the activity systems that develop in families are closely linked. Educators need to be aware of and plan for the messages that families potentially can take from educational sources. Differential access to these messages, and the occurrence of ambiguous and unintended messages are major concerns. The corollary of this, however, is that educators can have a substantial educative role with families.

> Educators need to be aware of and plan for the messages that families take from educational sources.

Researchers: Research is needed to investigate the properties of specific activity systems and how expertise develops within them.

Further Reading

Examples of reading activities analysed as tutorial systems:

Clay, M. M. & Cazden, C. B. (1990). 'A Vygotskian perspective on reading recovery'. In L. Moll (ed.), *Vygotsky and Education*. Cambridge University Press, Cambridge.

Ninio, A. & Bruner, J. S. (1978). 'The achievement and antecedents of labelling'. *Journal of Child Language*, 5, 5–15.

Descriptions of developmental outcomes:

Sulzby, E. & Teale, W. (1991). 'Emergent literacy'. In P. D. Pearson, R. Barr, M. L. Kamil & P. Mosenthal (eds.), *Handbook of Reading Research, Vol. 2*. Longman, New York.

Wagner, D. A. & Spratt, J. E. (1987). 'Cognitive consequences of contrasting pedagogies: The effects of Quranic pre-schooling in Morocco'. *Child Development*, 58, 1207–1219.

Research into caregivers' ideas:

Bornstein, M. H. (1991). 'Approaches to parenting in culture'. In M. H. Bornstein (ed.). *Cultural Approaches to Parenting*. Lawrence Erlbaum, Hillsdale, NJ.

Goodnow, J. J. & Collins, W. A. (1990). *Development According to Parents: The Nature, Sources, and Consequences of Parents' Ideas*. Lawrence Erlbaum, Hillsdale, NJ.

End of chapter notes

1. Not sharing a joint focus can arise from a number of sources. These include misunderstandings from different forms of communication, different pedagogies, and different experience with literacy activities reflecting the different cultural and social identities of teachers and learners. These forms of 'mismatch' are discussed in chapters 9 and 10. There is experimental evidence for the significance of establishing joint focus. One source comes from the highly successful intervention programme Reading Recovery (Clay & Cazden 1990), which deliberately attempts to develop shared goals in the initial stages of instruction.

2. The need to ask participants about the goals and ideas in activities is becoming more and more a necessary part of research into socialisation processes in general (Bornstein 1991; Goodnow & Collins 1990) and literacy in particular (Renshaw 1992). The framework of co-construction presented in this book assumes that ideas and goals are fundamental processes in learning and development.

3. Statements about long-term developmental outcomes need to be treated with considerable caution. Developmental processes are not like inoculations. A burst of stimulation at one point does not in and of itself ensure a particular product months later. Rather, the immediate development interacts with further processes over time. It is likely that the parents in Whitehurst *et al.*'s (1988) research continued their increased emphasis on book reading beyond the experimental period. Similarly, the children's increased vocabulary and expressive language may have enabled them to enter into more frequent and longer language exchanges with others around them.

4. Such comparisons are difficult to interpret. The observed differences may be due to differential experience with picture-books. Furthermore, they are almost certainly due to important differences in socialisation goals associated with membership in a particular societal or cultural group. Being a member of a particular social or cultural group may carry particular experiences and preferred ways of carrying out the activity. The contribution of cultural and social processes to differences between middle-class and lower socioeconomic-group families needs further clarification (e.g. Ninio 1980).

5. This phenomenon occurs during silent reading too. Stanovich (1992) has summarised research with children at school, showing that the amount of silent reading both in and out of school is closely linked to vocabulary growth and achievement in reading at school. This is a general finding for language development which applies across oral and written domains. A basic cognitive endowment of children is to seek connections and establish relationships. This leads children to actively construct words and their meanings from the ambient language in context (see McNaughton 1987).

6. A long tradition exists of analysis of the critical features of texts, and questions of what makes a written narrative or what makes an expository text are not simply answered. The descriptions and definitions which I have used here come from a number of sources (e.g. Pellegrini *et al.* 1990; Stein & Glenn, 1979). A recent book by Britton and Pellegrini (1990) contains extended discussion of the definition of narrative texts.

7. The relative merits of explicit and implicit teaching and learning were discussed in Chapter 4. It is important to note that the benefits of learning in which knowledge is revealed through the actions of the learner are determined by the nature of the activity and developmental needs. It is very efficient to teach letter knowledge directly in the service of developing expertise in other areas of reading and writing. As much of that knowledge is in the form of simple associations, direct instruction is very effective. Similarly, some concepts might need to be explicitly taught under remedial circumstances.

But the more that complex concepts and forms of expertise are deliberately focused on out of context, the greater the problems may be in achieving generalised and integrated ways of employing them in everyday contexts. On the other hand, teachers of English language need to identify structural and functional properties of language in order that students develop concepts to guide their reflective and generalised use of language (see Cazden 1993a and McNaughton 1987 for further discussion of circumstances and outcomes).

8. A similar argument about the restricted nature of some materials and activity structures can be made for beginning reading instruction. For example, the possibilities for incidental learning and personal systems developing is restricted in a highly regulated 'phonics' (sound-based) programme of instruction (see McNaughton 1987).

Chapter Six

The special case of reading storybooks

> **Focus**
>
> **Different ways of reading storybooks with children**
> - This chapter discusses why researchers and educators consider the activity of storybook reading to be important in the transition to school.
> - Activity systems for reading storybooks with children are described. For different systems the participants have different goals, different interaction patterns occur, and different tutorial configurations are present.
> - The forms of expertise that children develop from reading storybooks reflect the particular activity system that is experienced. Activity systems reflect and construct particular social and cultural meanings.

Alice was beginning to get very tired of sitting by her sister on the bank, and of having nothing to do; once or twice she had peeped into the book her sister was reading, but it had no pictures or conversations in it, 'and what is the use of a book', thought Alice, 'without pictures or conversations'.

Alice's Adventures in Wonderland, Lewis Carroll

I am good at prayers because, since I was a tiny boy, my father he get me and my brothers and my sisters and my cousins for to learn how to make the prayers. He also get us for to read from the Holy Book until now I am sixteen years old and am an expert in the reading of the Book.

Exam failure praying, Albert Wendt, 1986

Two forms of expertise in reading are captured in these two quotes. One is the form most often associated with storybook reading. The second is more often associated with learning Bible passages, but it occurs also with storybook reading.

The general features of these forms of expertise were introduced in the previous chapter. In this chapter I explore in greater depth their characteristics, the social and cultural meanings with which they are associated, and how families might learn new ways of reading storybooks.

Reading for Narratives: A Closer Look

Developmental significance

> Reading storybooks with children before they start school has come to be seen as an especially significant literacy activity.

Reading storybooks with children before they start school has come to be seen as an especially significant literacy activity. It is a core component of many intervention programmes for families with young children 'at risk' in terms of achievement at school (see Fox 1990). In New Zealand it is part of the Parents as First Teachers Programme (an intervention programme based on a model developed in the United States (Meyerhoff & White 1986) and has been present in advice to new parents provided nationally in postnatal check-ups for many years (Chapter 2).

> Children in families that provide substantial early-literacy experience, including reading books with them, do relatively well in beginning reading.

The evidence for the significance of storybook reading before school is generally quite consistent. Families that provide substantial early-literacy experience, including reading books with their children, have children who do relatively well in beginning reading. In Western industrialised schooling systems this relationship can be found in different social groups and across ethnic groups, although there are important factors which influence this connection, including relationships between communities and their schools (e.g. McNaughton & Ka'ai 1990; Snow *et al.* 1991).

More specifically, several studies have found significant relationships between the frequency of being read to at home and success in beginning reading at school (see Sulzby & Teale 1991 for a review). The effects can be quite specific. In two studies of English children Wells (1985) found this relationship between reading achievement after two years at school (in reading comprehension as well as reading vocabulary) and the amount of shared storybook reading at three years. Other literacy events, such as just looking through picture-books and magazines, did not have this relationship. It is frequent, progressive experience with the social interaction and language that surrounds the reading of storybooks that seems essential to the relationship.

> Frequent, progressive experience with the social interaction and language that surrounds the reading of storybooks seems essential to the relationship between being read to at home and success in beginning reading at school.

What these studies provide is evidence of specific developmental outcomes of the activity. The question then becomes, how does storybook reading contribute to this? The answer comes in several parts. The relationships are found with storybook reading as a particular activity, defined by particular goals, interaction patterns, and a tutorial configuration. The relationship is brought about by particular forms of expertise that develop within the activity. Finally, the

> The developmental association with achievement at school occurs because important relationships exist between the activity at home and complementary activities at school.

developmental association with achievement at school occurs because there are important relationships between the activity at home and complementary activities at school. Each of these will be identified.

Collaborative participation: pedagogy and culture

Recently, Sulzby and Teale (1991) reviewed a number of intensive descriptions of reading for narrative meanings. I now take a closer look at a study that Gwenneth Phillips and I carried out (Phillips & McNaughton 1990) which is representative of the general findings summarised by Sulzby and Teale (1991). Its likeness underscores their conclusion:

> In the situations and cultures studied, children almost never encounter an oral rendering of the text of the book in a storybook-reading situation. Instead, the words of the author are surrounded by the language of the adult reader and the child(ren) and the social interaction among them. During this interaction the participants cooperatively seek to negotiate meaning through verbal and nonverbal means.
> (Sulzby & Teale 1991, p. 733)

The descriptions in the 1990 study came from the homes of 10 families who identified themselves as interested in books and book reading. The families were Pakeha and the incomes and occupations of family members placed them in the top two socioeconomic groups in New Zealand. Each family comprised two adults and one of two children, and the mother was the major caregiver. The children (six boys and four girls) were aged between 3 years and 4 years 6 months.

Through diary records and indirect observation it was established that these families were 'book oriented'. For example, someone read frequently with the preschooler (a mean of 87 books read over a 28-day period); and selectively (95 per cent of the books read were books with a narrative structure). The families estimated that they possessed 450 books (although their estimates ranged from 50 to 500). Of the 450 books, 300 were children's books. The deployment of resources, including the selection of reading materials and time commitment, created a Zone of Promoted Actions (see Chapter 2, p. 17), in this case a 'zone' of frequent book-reading sessions. Both readers (predominantly the mothers) and the children initiated these sessions but it was primarily the child who chose the book to be read.

We gave new storybooks to these families to read, which they read over the following month. These storybooks were specially chosen as being unfamiliar but they were rated by the families as similar in type and value to their usual books. This enabled us to examine tutorial processes on texts which had the same degree of familiarity across the group of families but were read as typical sorts of storybooks. The first, second, and last readings were audio taped and subsequently analysed. Families were asked to read these texts as they usually read storybooks.

In keeping with Sulzby and Teale's description above, conversations that punctuated the reading were very obvious. The goals of the reader and the child in storybook reading were established through an analysis of interactions inserted into the reading, either by the adult reader or by the child. Analyses of these exchanges (from the point at which reading the text stopped to when it started again) revealed that both the readers and the children were focused, almost exclusively on the narrative structure of the texts they read, rather than on, for example, concepts about print or unrelated descriptions of illustrations (see Table 6.1 below).

Table 6.1 Focus of interactions in storybook reading: mean percentages of exchanges (647) for 10 families

Focus	Mean per cent
Narrative	86%
Print	3%
Book /Other	7%
Other	3%

Of the 647 exchanges analysed, 86 per cent were focused on the narrative. A very small percentage were focused on print conventions (e.g. identifying letters), or on other facets of the book that did not relate to the story (e.g. counting the pages). Other exchanges (3 per cent) were not able to be coded or involved child management.

Narrative interactions were described in Chapter 5 (pp. 90–91). They begin with comments, directives, or questions which are focused on information relevant to, or consistent with, the events and the goals of the narrative. The interaction patterns found in this study were characteristic of the type of tutorial I have called collaborative participation (see Chapter 4, p. 69). Reader and child collaborated in these exchanges to construct meanings for the text. The preschooler participated as a full conversational partner in setting the topic. (The readers and the children initiated exchanges at similar average rates of 3.8 and 3.7 per bookreading respectively, with most of these focused on narrative meanings.)

The properties of the guidance provided by the collaborative-participation tutorial (see Chapter 4, pp. 64–68) are illustrated in excerpts from two families reading the same text, *What Made Tiddalik Laugh?* (Troughton 1977). One of these excerpts was introduced on page 69. It is reproduced on page 107, with an example from a second family. In both cases exchanges developed at the same point, at the beginning of the text where the narrative problem is set (the giant frog has drunk all the water and the other animals have to find a way to get it back).

In the initial reading the mother in the first family highlighted the narrative problem by clarifying what would happen to the animals. After about four weeks of reading (later reading) she gave less direct support. By reducing her direct support she increased the instructional pressure, probing her child's awareness of the event. The exchange ended with the child representing the meaning which they had shared on earlier readings and upon which they had agreed (there would be nothing left and no water to drink). In so doing she identified the narrative problem, which the mother affirmed by extending the child's meaning (*'And they won't be able to have a swim, will they?'*).

Sample dialogue 6.1
Example of collaborative participation

Reading a storybook

TEXT   In the dream time there lived a giant frog called Tiddalik. One morning when he awoke, he said to himself, 'I am so thirsty, I could drink a lake!' And that is what he did.

Family One
(Initial reading)
MOTHER   My word! There's nothing left for the birds and the turtles …
CHILD   No.
MOTHER   … and the lizards, is there?
CHILD   No.

(Second reading)
MOTHER   Look at him drinking all the water out of the lake. None of the animals have got any water left.
CHILD   No.

(Later reading)
MOTHER   What's he doing?
CHILD   He's drinking up a pond.
MOTHER   Yeah. The lake. All the water in the lake.
CHILD   Yeah.
MOTHER   Poor animals and birds.
CHILD   Yes. They will have no water to drink.
MOTHER   No.
CHILD   Because they're thirsty.
MOTHER   And they won't be able to have a swim, will they?

Family Two
(Initial reading)
READER   Goodness! He's a big frog.
CHILD   And look at all those little frogs.
READER   Yes [laughing]. They're all looking at him. Probably going, 'Huuu. What an enormous one!'

(Later reading)
CHILD   I wonder what those frogs are looking at him for?
READER   Why do you think?
CHILD   I don't know.
READER   What! Do you think they're a bit surprised at how large he is and what he's doing?
CHILD   I think … well … he's eating all their water.
READER   Yes, I do too.

A similar shift occurred in the exchanges in the second family. After several readings the child adopted the mother's interrogative role and her focus in the earlier reading. As with the first family the child took over more of the task of clarifying the problem. For these two families the increasing familiarity provided by the format of the joint activity enabled greater participation. For other families this increasing expertise and familiarity was associated with less overt participation. This occurred as children developed personalised forms of the joint co-construction. A shift at some point to less collaboration, is found in other research reports, too (Sulzby & Teale 1991).

Analyses of the way in which all 10 families read the books over the four-week period showed that the readers concentrated on clarifying the narrative on the first readings. Over time, however, they reduced this emphasis as children more often initiated exchanges in which they tried to clarify what was being read. The reader meanwhile shifted the focus towards making links with anticipated text segments. It was apparent that the children were taking more and more responsibility for the comprehension to the extent that they initiated questioning and checking. Self-initiated questioning and checking by the children provided evidence for the development of self-regulation (e.g. Chapter 3, p. 46).

These exchanges have pedagogical power and what this enables children to be able to do is described in greater detail below. But it is important also to identify what messages are carried in the activity about identity: who one is and how one becomes an expert member of the family. After all, a major claim in the co-constructivist explanation is that social and cultural messages are reflected and constructed in the literacy practices. Among the complex interwoven features of this reading activity are two features which can be contrasted with other ways of carrying out the activity — the authority of the text, and the role of the individual. But in making these contrasts one qualification needs to be kept in mind. Isolating the cultural meanings which I discuss (for example, about the role of the individual) inevitably simplifies their significance in the overall environment of family activities.

> Collaborative storybook reading carries messages about who one is and how to be an expert in this family activity.

Authority, negotiation, and language

I have already pointed out that the majority of the exchanges in the study focused on the narrative. Considerable time was spent clarifying, anticipating, or integrating shared meanings within the narrative. In this sense, the language of the text itself, carried in its linguistic structure, was an authority. The language of the text held meanings which were elaborated upon and discussed. But possible meanings from the text could be checked against the authority of the language in the text.

> One message is that the language of the text is an authority. It holds meanings which can be elaborated upon and discussed. Possible meanings can be checked against what has been written in the text.

A telling example of this occurred in an exchange when a mother and her child first read *What Made Tiddalik Laugh?* (p. 109). In discussing the final illustration the child questioned an assumption her mother had made. Her mother had assumed there were two platypuses in the story ('*... and where's the other one? Can you see the other one?*'). This assumption had come from the misinterpretation of an earlier ambiguous illustration (p. 109). The earlier illustration is of the platypus's tunnel; the picture is spread across two pages and shows the platypus in two positions. The mother had assumed from this illustration that there were two platypuses in the story that lived together.

The child negotiated this meaning, appealing to her memory of the text ('*But it said ...*'). Finally the mother agreed. But only after they had checked the illustration and she referred to the accompanying text. That the child could question her mother's understanding and force her back to the text expresses significant cultural and social messages.

THE SPECIAL CASE OF READING STORYBOOKS

Sample dialogue 6.2
Negotiation of meanings

Reading a storybook
TEXT   *Thank you for making Tiddalik laugh.*
CHILD   He's still laughing.
MOTHER Mmm. And where's the platypus and where's the other one? Can you see the other one?
CHILD   Yep — no.
MOTHER That's a wombat isn't it?
CHILD   Is that a wombat? He's got wings.
MOTHER No, a wombat doesn't have wings.
CHILD   Platypus.
MOTHER Yes, there's another one in the water.
CHILD   But she lives by herself.
MOTHER No, I think there are two of them. There are two of them down in the tunnel.
CHILD   No, there's not.
MOTHER You look at the picture, down in the tunnel, there's two of them.
CHILD   But it said there was one! Mum?
MOTHER There's more than one platypus.
CHILD   They don't live together [inaudible] … says she had no one. Yes.
MOTHER [reading] *… she kept herself to herself.*

The pages from the storybook *What Made Tiddalik Laugh?*, which were the source of the negotiation. The double-page layout suggests two platypuses, while the text refers to only one

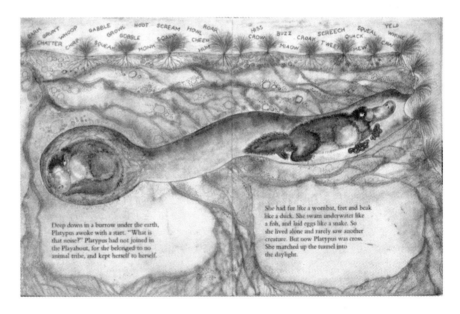

Gee (1990) has summarised at least one dimension to the cultural messages involved in these sorts of exchanges. It refers to the emphasis on examining possible meanings in what has been written using the information in the sentences; this is illustrated by the child in Sample dialogue 6.2 searching and confirming meanings. Gee says:

> Anglo-Canadian and American mainstream culture has adopted a model of literacy based on values of the essayist prose style, that is highly compatible with modern

consciousness. In essayist prose, the important relationships to be signalled are those between sentence and sentence, not between speakers, nor those between sentence and speaker. For a reader, this requires a constant monitoring of grammatical and lexical information. With the heightened emphasis on truth value rather than social or rhetorical conditions, comes the necessity to be explicit about logical implications. (p. 63)

The role of the individual

> The focus and the ways of participating in storybook reading may be associated with another set of meanings about the role of the individual that identify literacy as an individual possession.

The focus and the participation structure of the *'Tiddalik'* exchanges are associated with another set of meanings about the role of the individual. Gee also has discussed these meanings, which are associated with middle-class forms of literacy. They identify literacy as an individual possession. The message constructed in these exchanges is one which values the individual gaining or possessing knowledge.[1] This complements the previous messages about authority, negotiation, and language. These focus on the individual's close attention to meanings conveyed in the text, and on the logical implications in what is being read.

This emphasis on individual negotiation and possession can be detected in the shared control of the exchanges that occur during reading in white middle-class families. The child's status in this has been labelled as that of 'co-operative negotiator' (Heath 1983). This pattern is a pervasive feature of child rearing by white middle-class families, being present in early oral exchanges between children and their caregivers.[2]

Yet another vehicle for meanings about the role of the individual is provided in participation structures. In the Phillips and McNaughton (1990) study of storybook reading in middle-class Pakeha families, diary records over a month showed that most of the families with more than two preschool children still organised reading sessions on a dyadic basis. This emphasis on dyadic activity also added to the messages carried about authority and the role of the individual in the interaction.

Forms of Expertise and their Variations

> From collaborative storybook reading children develop a knowledge base for how written stories work.

Children develop particular sorts of expertise from collaborative storybook reading. Their expertise includes a developing knowledge base for how written stories work. This knowledge can be tested by an adult asking a child to read a familiar book, but without helping the child (see Sulzby & Teale, 1991). From the child's retelling of the story, judgements are made about what the child knows about written language in books. Within this activity, children's understanding can be shown to shift from the earlier picture-book strategies of labelling, to creating their own oral narrative based on the illustrations, to constructing a story with the properties of written language, including intonation patterns, syntax, and vocabulary. Ultimately children come to read familiar books conventionally, that is, using the written words.

These generalised ways of knowing about storybooks are based on the child's developing familiarity with the language found in books. They include knowledge about how information can be found in texts (such as searching for confirmation in Sample dialogue 6.2), and knowledge of how books contain meanings. Other closely related sorts of knowledge have been described by researchers, too. They include familiarity and knowledge of the units of language such as phonemes, words and syntax, and the concepts about print described in the previous chapter (see Dickinson *et al.* 1992).

But these forms of knowledge are not necessarily inevitable outcomes of reading books to children. They are the outcomes of a particular activity system. For example, children's knowledge of how written narratives work and their strategies for comprehending the meanings in books arise from collaborative participation focused on the narrative. The expertise is associated with the degree to which extended and elaborated discussion focused on events in the story takes place (Dickinson *et al.* 1992).

> Children's knowledge of how written narratives work and their strategies for comprehending meanings in books depends on the degree to which extended discussion focused on the story takes place.

Researchers have described differences between families in the patterns of interaction which I have described here as collaborative participation. These researchers' descriptions use the storybook-reading activity as it has been described here as the standard. Where differences have been found in interaction patterns across sociocultural groups, they have been described as variations from the standard form. There are some methodological problems in these studies which have to do with familiarity and expectations. But in studies of low-income families where storybook reading has been a regular family activity, families are more often described as interacting with a more limited focus on extended narrative meanings and with less co-operative participatory dialogue than middle-class families (Heath 1982; Sulzby & Teale 1991).[3]

One major implication follows from an uncritical acceptance of these comparisons. It is that there is a right way to read storybooks to children. Scaffolded storybook reading, described in terms of both a focus on narrative meanings and in terms of a collaborative style, is assumed to be the appropriate and most effective way to socialise children into expertise with written language. What follows is the conclusion that those families who do not read with their children in this manner are socialising their children inadequately.

The reasons for this view include the relationships between storybook reading and school success which were noted at the start of the chapter. The expertise that develops within the activity of reading for narrative provides a reason why children who have substantial experience tend to achieve well in school reading, continuing past the initial stages of instruction. On the one hand, school literacy activities recognise and build on the expertise, and on the other hand, children are able to make more sense out of the activities they encounter (see Heath 1982; Snow 1992; Gee 1990).

> Differences between social groups indicate an activity with different social and cultural meanings.

The detailed descriptions of reading storybooks reviewed by Sulzby and Teale (1991) are based on intensive case studies and group studies, mostly of white middle-class families. But several studies have found the presence of substantially different patterns from those I have just described. Rather than being a matter of less effective or more limited use of the activity, the extent and nature of those differences indicate a very different activity — an activity that has different social and cultural meanings for the participants.

Pedagogical Dexterity

The 17 families with whom we worked in the SOL study (p. 19) read books with their children. Book reading was a daily occurrence and storybook reading, in particular, was frequent. The resulting activity looked like the standard activity associated with progress at school. This is not surprising because all of the families were selected as having children who had progressed well at school. We subsequently found that generally the 4-year-olds also made good progress.

In the Maori and Samoan households, triadic settings for book reading were typical, and other members of the family, such as an older sibling or an auntie still at school, often took the role of reader. As with the other families, samples of storybook reading revealed the presence of the collaborative participation tutorial pattern. The majority of insertions (60 per cent of total insertions) and subsequent exchanges showed that readers and children collaborated in constructing text meanings.

But on some occasions a different form of interaction took place.

Performance routines occurred when part of the text was repeated by the child. Imitations often were signalled by the reader through verbal and paralinguistic (language-related) means, such as pausing and a final rising intonation.

On these occasions (a mean of 16 per cent of the exchanges) the reader indicated that a model had been provided and the child's task was to imitate. The performance routines were particularly noticeable on texts that were not storybooks. Examples of both types of interactions are shown in Sample dialogue 6.3. They come from one book-reading session in a Maori family. A storybook was read collaboratively. We found that all 13 insertions had a narrative focus (one of which is reproduced — *'What are they doing?'*). However, a beginning reading text bought home from school by a cousin was read using performance routines (performance routines occurred on every line of the text).

> In Maori and Samoan households in the SOL study, triadic settings for book reading were typical, and other members of the family, such as an older sibling or an auntie still at school, often took the role of reader.

Sample dialogue 6.3
Shifting tutorial styles in one book-reading session by one family (child aged 4 years 11 months)

(Book one — home storybook)
TEXT   *Patch had been given the job of painting white lines for the running lanes.*
READER What are they doing there D... ?
CHILD   Painting ... a line.
READER So that they can run down the track straight.

(Book two — cousin's school book)
TEXT   *Andrew had an engine called Red Streak.*
READER Andrew.
CHILD   Andrew.
READER had.
CHILD   had.
READER an engine.
CHILD   an engine.
READER called.
CHILD   called.

In Chapter 5 these directed-performance tutorials were described as having a different focus. The goal of the activity of Reading for Performance (p. 92) is an accurate rendering of the text, and it was not overtly collaborative. The changes in storybook reading showed that the Maori and Samoan families could read with textual (or perhaps pedagogical) dexterity. They selected a wider range of books than the 'mainstream' counterpart families described by Phillips and McNaughton (1990), and could shift the type of tutorial within and across texts.

Reading for Performance

The performance style of reading is particularly obvious in transcripts from a group of eight Tongan families living in New Zealand (pp. 31–32). The book-reading settings were all multiparty but with an adult caregiver present, and in some families someone other than a parent, such as young aunts and older siblings, assumed the role of reader.[4]

Three 'stages' of tutorial which employed only performance routines were observed. Samples are shown in Sample dialogue 6.4. In the initial stage the reader demonstrated a part of the text and the child imitated. In the second stage, the child read much of the text alone and hesitations or misreadings were repaired by the reader. In the third stage the whole text was performed alone. Research has revealed instances where preschoolers have read texts entirely by themselves.

Sample dialogue 6.4
Examples of Performance routines

(Early reading)
READER   Why hares have long ears.
CHILD   Why hares have long ears.
READER   Once upon a time.
CHILD   Once upon a time.
(Familiar reading)
CHILD   She is up on the tree.
        She is up on the … [pause]
READER   Stool.
CHILD   Stool.
        She is up on the horse.
(Multiple reading)
CHILD   One day mouse, rabbit, and elephant went to the fair. Mouse went up, rabbit went up, elephant went up.

Performance tutorials: pedagogy and culture

In New Zealand some families practise reading for performance relatively exclusively, even with storybooks. Others are pedagogically dexterous, shifting between two major forms. In the latter case these families choose to use an activity system which is a 'non-standard' form.

This tutorial configuration produces an activity system in its own right — Reading for Performance. All the hallmarks of a tutorial configuration are

> Some families read storybooks for performance, which creates a particular activity system.

present when the activity is carried out effectively. For example, the support can shift over time creating a tutorial structure which is adjustable and temporary. At the beginnings of learning the complete demonstration of segments of a new text is common. This support gradually reduces as systematic modifications to the amount of text which is demonstrated take place. Models for words or segments of texts can be provided contingently on the child's growing control over the text.

Within this structure, responsibility for accurate performance shifts to the child. The learner comes to repeat and even repair performance by themselves. It is possible to detect the shared focus characteristic of effective tutorials. It can be detected by the lack of confusion over what is expected, for example in the fluidity of the repairs, such as in the sample on page 113 (*'She is up on the … Stool … Stool'*). It can be detected also by the easy participation of siblings and other family members, and by the finding that some families could shift, seemingly with ease, between the two forms of tutorial.

Does the self-regulation which is associated with collaborative participation develop through this non-standard tutorial form? This is a thorny theoretical question because the behavioural indicators of self-regulation in mainstream book reading — checking, reflection, and inquiry — do not appear to be appropriate indicators of the performance activity here. This reduced need for overt self-correction, inventive exercise, and play with patterns has been described in other non-literacy activities in other cultures too.[5]

> Becoming an expert performer requires monitoring to prevent inaccuracy and, at least during acquisition, limiting generalisation.

In the case of becoming an expert performer there are important responsibilities — to monitor to prevent inaccuracy and, at least during acquisition, to limit generalisation. What this means is that the need for overt checking is reduced through the anticipation and prevention of possible mistakes and automatic production. But this also suggests that the self-reflection associated with the collaborative-participation tutorial is itself a form of valued expertise. It serves particular needs, such as the need to be able to generalise and engage in reflective enquiry.

In the activity of Reading for Performance the children were engaged in a form of a mature task which expressed and constructed social and cultural meanings. Among other things this included the need for accurate performance in cultural and spiritual activities. Messages relating to the themes I discussed for collaborative participation in this activity can be identified (p. 108), namely the nature of authority and the role of the individual.[6]

Authority

> The performance tutorial is a core part of Maori preferred pedagogy. Traditional key instructional devices were rote learning and memorisation, especially for essential shared knowledge.

The performance tutorial form is a core part of Maori preferred pedagogy (Metge 1984). In traditional contexts of formal teaching the key instructional devices were rote learning and memorisation, especially for the essential shared knowledge of the tribe contained in genealogies, songs, and narratives.

> Maori puukenga (knowledgeable experts) set unfashionable store by memorisation and rote learning. The memorisation process itself has become a matter of 'teaching' in the form of group practice led by an expert. The puukenga leading

these learning sessions follow a similar procedure, with variations, repeating each name or phrase a couple of times, adding the next, repeating again from the beginning, adding the next, repeating again, and so on to the end. Some 'teachers' using this method give little or no explanation of the content until the words have been mastered; others discuss meaning at intervals during the memorising, arguing that it is easier to remember what is understood. … it is clear that rote learning is not an end in itself but the first step towards the goal of meaningful performance. … knowing which to use and when, and that depends on knowing the background, being able to size up the situation, and to make the right choice. (Metge 1984, p. 8)

Among other things this pedagogy described by Metge can be linked to beliefs about the authority of oral texts and the nature of knowledge. In an oral tradition cultural needs for the accuracy of oral texts and a stress on the preservation of knowledge are central concerns. It is also linked to a principle that knowledge is precious and to be treasured. This sense of guardianship and protection reinforces the value of representing knowledge accurately and without embellishment.[7]

> In an oral tradition cultural needs for the accuracy of oral texts and a stress on the preservation of knowledge are central concerns.

For Samoan families, written texts reflect a set of values associated with the authority of church teachings and a strong respect for the church and for schooling. Early literacy experiences and resources are promoted and channelled by the pastor school and the church. Because of the relationship between the church, schooling, and literacy, the reading of texts also represents the authority of elders, and supports adherence to significant religious beliefs (Duranti & Ochs 1986).

These and other associated values can be seen in daily readings from the Bible and in family lotu, which are common in Samoan families (McNaughton & Ka'ai 1990). Several purposes are entailed in this activity. There are the purposes associated with religious beliefs, but others include the cohesion of the family and its commitment to shared beliefs, roles, and responsibilities in Fa'a Samoa. Earlier I described a Samoan family's lotu at which grandparents, parents, children, and grandchildren were present (see Chapter 2). The performance tutorial was used for teaching biblical passages. The patterns of participation carried important messages about nurturance and social responsibility, too, as well as the authority of the text and the person leading the devotions.[8]

> For Samoan and for Tongan families the performance of written texts reflects values about the authority of literacy associated with church, schooling, and family.

The authority of texts and the occurrence of performance tutorials are associated with religious values in the eight Tongan families described on page 31. For example, the performance tutorial is a standard form used in Sunday school to teach hymns and church texts. The Tongan children from these families were observed learning hymns, Bible verses, and articles of faith at their Sunday school, where performance tutorials were used. A performance tutorial segment part way through the learning of a new hymn is described in Chapter 10 (p. 183). Similar cultural messages were being transmitted across activities at home and at church in the use of written texts (Bible verses, hymns, etc.), again creating multiple messages for the process of socialisation.

The role of the individual

As with the activity in Pakeha middle-class families, the participation structure in performance-directed tutorials carries important messages about the nature of literacy and individual responsibility. Participation differences that have occurred in several studies were noted in Chapter 2 (p. 32). Multiparty sessions tended to occur in the Maori, Samoan, and Tongan families. And these represent, at least to some degree, arrangements by choice. The resources of these families were deployed in the participation structures to achieve family goals.

The preference for group learning expressed by Maori (Metge 1984; Smith 1987) often has been interpreted by non-Maori writers as being exclusive of personalised learning interactions. Certainly individual instruction which focuses solely on the learner and the possession of knowledge is problematic. In descriptions of preferred pedagogy teachers are reluctant to single out an individual for verbal praise or blame in public for fear of making them whakahihi (conceited) or whakama (so embarrassed that they retreat into themselves) and so disturb their relations with the rest of the group (see McNaughton & Ka'ai 1990; Metge 1984).

Extended personalised interactions have been observed, however, in group and shared activities in Maori educational settings (e.g. Hohepa *et al.* 1992). The important feature associated with an individual's learning here is that the development of expertise carries responsibilities for the group. For Maori, knowledge is a group possession, not belonging to the individual, and to be used in the service of the group.

This pedagogical preference can be traced to deeply held values associated with whanaungatanga, and the tuakana-teina relationship is one expression of this (see Chapter 2, p. 32). Multiparty settings extend the meanings of these two concepts and reinforce the relationship between the individual and the group.

Tutorial Configurations Revisited

The presence of at least two major configurations of tutorials (collaborative participation and performance-directed) in the activity of storybook reading illustrates how emergent literacy is rooted in social and cultural lives. The analysis in this and the previous chapter has assumed that there are general properties of socialisation tutorials which are applied in different configurations under different sociocultural circumstances.[9]

The qualification about tutorials I noted in Chapter 4 (in note 8) needs to be reiterated here. It relates to the concern raised by Van der Veer and Valsiner (in press). They describe a 'blind spot' which has followed from a misapplication of sociocultural concepts in educational writings. It is the uncritical assumption that an 'educational utopia' is achievable with collaborative encounters. Clearly, good and poor tutorials are possible within multiple socialisation goals. The task facing researchers and educators is to plot how variations in configurations facilitate effective learning. It is to plot also the ecological conditions that enable families to function with effective tutorials, and identify those that impede them.

Different configurations have their own internal criteria for effectiveness. Tutorials can be well or poorly implemented to achieve the purposes of the family's literacy practices and more general socialisation goals. This applies to performance-directed as much as to collaborative participation tutorials. For example, the degree of overt collaboration and the coherence of the questions and comments in collaborative participation might determine the development of comprehension strategies for narrative texts. In performance-directed tutorials the clarity and chunking of the model may determine the speed of acquiring accurate performance.

Raising the Standard

A comment needs to be made at this point. It concerns the acceptance of a view that collaborative participation is the appropriate configuration for tutorials for storybook reading. Tutorials take place in activities, and activities reflect, express, and construct social and cultural practices. One reason why some configurations may have been judged as inadequate is because researchers have not understood the relationships between what is being observed and their social and cultural meanings.

> Configurations may be judged as inadequate because researchers do not understand the relationship between what is being observed and their social and cultural meanings.

Recent evidence suggests that published research involving African Americans has substantially decreased in journals of the American Psychological Association over the last 20 years (Graham 1992). In New Zealand similar marginalisation and ethnocentricity in research has occurred and the small number of developmental psychologists who are Maori, or members of Pacific Island groups are faced with difficult barriers of resourcing, editorial policies, and research agendas (Cram 1993).[10]

Another reason for the normative view stems from assumptions about the nature of development. Literacy development often has been assumed to follow a fixed single sequence dictated by universal stages of cognitive development (e.g. Goodman 1990) or constructed through a core set of concepts (e.g. Mason & Allen 1986). In either case development is seen as moving inexorably towards a final state defined by schooled forms of literacy. Different configurations from the standard scaffolding model of collaborative participation have been linked to the further development of literacy at school. Limited storybook reading, as well as infrequent experience of storybook reading, has been associated with problems in the development of literacy at school (Sulzby & Teale 1991).

> The normative view stems from assumptions about the nature of development. Literacy development often has been assumed to follow a fixed single sequence dictated by universal stages of cognitive development, as moving inexorably towards a final state defined by schooled forms of literacy.

This position means that diversity in book-reading experiences comes to be seen explicitly or implicitly as non functional or inadequate. An alternative view, however, from which the concept of configurations is derived, sees literacy in terms of increasing participation in social and cultural practices, within which multiple forms of expertise may develop. In this view diversity is expected and seen to have important social and cultural functions.

But there are problems in an uncritical celebration of pedagogical diversity. Developmental trajectories associated with different pedagogies intersect with developmental sites other than the family. Some forms of literacy carry more

> There are problems in an uncritical celebration of pedagogical diversity.

cultural capital in an educational system than others. For example, beginning-reading instruction in schools in New Zealand recognises expertise in reading texts for narrative meanings, but may not recognise expertise in recitation.

The challenge here is to employ explanations which clearly locate expertise in terms of the cultural identities of families and the immediate and more distant sites for their children's development. The analysis is limited if multiple trajectories and multiple settings are not considered (see also Damon 1991). This is especially so at the transition point between home and school practices of literacy where processes of cultural power and the 'privileging' of different cultural messages occurs. For example, schools may be more likely to recognise and build upon expertise arising from collaborative-participation tutorials.

> Expertise needs to be located in terms of the cultural identities of families and the immediate and more distant sites for their children's development.

Dexterity and Parent Education

> Given the educational significance of storybook reading for narrative, what are the implications for optimising literacy development and educational achievement?

Given the educational significance of storybook reading for narrative what are the implications for optimising literacy development and educational achievement? There are three possible solutions to such an issue. One solution is to change the selection of activities at home and parental patterns that are not like the standard activities and configurations. Another is to change school literacy practices to be responsive to different activity systems and their associated forms of expertise. A third solution is to do both. The last option may be the ideal, and there are examples of optimisation programmes that have attempted to create a collaborative enterprise in which community and school influence each other's literacy practices (Heath 1982).

In this section I discuss this ideal in terms of the impact on, and educational processes for, families. The theoretical framework I have developed claims that expertise in literacy cannot be divorced from our identity. Altering literacy practices, for example, by introducing a new activity, will alter social and cultural functioning in some way. Clearly, there are significant issues to do with control and power at work in such an enterprise. Simply trying to replace social and cultural messages is a form of, at worst, colonialism and, at best, assimilation. The solutions to this dilemma lie in the avoidance of notions of replacement. The dexterity of the families described in the SOL study serves as a model for how equitable yet developmentally functional forms of expertise might be added.

Family education and learning to read for narrative

Is it possible to learn to read storybooks for narrative when one has not done this, when one has not experienced this and it presents a substantially different socialisation activity? Among other things are issues here of deployment of resources, including time, space, and materials, which are by no means minor considerations (see Chapter 2, p. 18). But at the level of family education two sorts of collaborative programme are possible and suitable for different groups of parents.

The storybook as a narrative prop

Storybooks have been introduced to families under circumstances where storybook reading has not been a familiar regular activity. Heath and Branscombe collaborated with a young African American mother, who was described as unemployed and having dropped out of high school, and unaccustomed to reading books (see Heath & Branscombe 1986). The introducing book reading for her 2-year-old son was a project for school credit. The aim was to read 10 minutes per day and to tape-record the sessions. Branscombe visited with books (e.g. alphabet books and storybooks) and reviewed tape recordings, and the three participants (parent, teacher, and researcher) wrote to one another focusing on the language project.

Initially the routines during book reading were mixtures of performance routines for words ('*Say [X]*'), or requests to repeat sections of texts or labelling routines ('*Look at [X]*'). Very infrequently expansions and commentaries on the topic occurred. Exchanges were primarily under the mother's control and relatively non collaborative. In this activity the 2-year-old's language and participation changed. He began to tell stories, both fictional and factual. He began to collect plots, dialogue, and characters from books and other sources. He became able to sustain the topic of a narrative and understand characterisation. What is most significant to the development of an effective system, his mother began to respond to these developments within interactions. This shows that a personalised system began to influence the dialogue patterns during storybook reading.

The conditions for the effective and sustained development of reading for narrative cannot be ascertained from this study. But the features apparent in the case study included the development of joint focus (i.e. collaboration) between researcher, teacher, and parent. They included also the meaningful task of the project, and feedback and support. Finally, the resources of books and long-term collaboration in a context of mutual support also would have been important ingredients.

Direct teaching

Heath's programme could be described as a tutorial with the mother that itself was like collaborative participation. The complex narrative exchanges and the dynamic tutorial emerged, or, using Cazden's (1993) terminology, were revealed within the structure over time. But there is evidence that these patterns are unlikely to develop where parents' ideas about pedagogy and goals for book reading are not focused on the narrative meanings. The educational significance of parents' ideas in emergent literacy includes the relationship between their ideas and the availability of particular literacy activities in the home. Gallimore and Goldenberg (1993) link the limitations of introducing Spanish storybooks into lower-socioeconomic, Spanish-speaking families with kindergarten children to parental beliefs. Although storybook reading with children increased, it did not include parent–child interactions focused on text meanings.

The reason for the ineffectiveness of introducing storybooks was that the parents believed that the goal was correct performance of the task as homework, and they focused on the accurate rendition of text segments. This interpretation of the significance of beliefs is supported in a study by Renshaw (1992). He showed that parents' interpretations of researchers' requests to read to their children were related to how they then read. More direct forms of optimisation may therefore need to be outlined in which the routines are specified and the tutorial structure explained. This may be necessary where culturally preferred patterns exist, such as those used by the Tongan parents.

It is extremely important that these sorts of programmes remain collaborative. Where direct instruction has been used as an intervention with parents, where there has been little collaboration and mutuality, programmes tend to be short lived and rejected (Fox 1990). There is every indication that it is relatively easy to learn a new activity given appropriate conditions (Sulzby & Teale 1991). These conditions include using the family's preferred learning styles, appropriate role models, resourcing, and long-term collaboration. In the example of the Tongan families this might mean: working with Tongan families who are familiar with performance-directed tutorials; incorporating this form into the teaching; working with Tongan educators; providing appropriate books (in Tongan and English) as well as access to these; and working in ways that enable the families to influence the educational process and outcomes.

Dickinson and others have defined for their research purposes narrative sorts of talk during storybook reading, which are related to later progress in reading (particularly for comprehension) at school. These forms of talk can be used as models for types of exchanges which can be taught to parents. For example, Dickinson et al. (1992) provide a definition of 'non-immediate utterances' as the general category of exchanges associated with the development of narrative knowledge and decontextualised language. These exchanges 'move away from what can be seen on the page and include thoughts on and analyses of character behaviour or motivation, discussion of vocabulary, or connections between the story and the child's own world. [A non-immediate utterance] includes feedback and requests for clarification of meaning' (Dickinson et al. 1992).

Their definitions are similar to ones I have used in the research studies described in this chapter. In these studies it has been useful to distinguish between exchanges which clarify (extend the immediate text), integrate (what is read with what has been read), and anticipate (forthcoming events and characters).

> There is every indication that it is relatively easy for families to learn a new activity form given appropriate conditions, which include using the family's preferred learning styles, appropriate role models, resourcing, and long-term collaboration.

Summary

Each of the propositions on which the previous chapter was based guided the discussion in the present chapter, but I concentrated on just one activity rather than describing several activities. Overwhelmingly, it has been the activity of storybook reading that has fascinated researchers and educators. Propositions Three, Four, Five, and Six summarise this chapter.

Proposition Three: Literacy practices are expressed in specific activities which have identifiable constituents. These include goals, rules for participation, and ways of carrying out the activity. This chapter has examined the activity of storybook reading

> Systems for learning and development take form within the activity of storybook reading.

Proposition Four: Systems for learning and development take form within the activity of storybook reading as a product of the child's actions and the actions of significant others.

Proposition Five: Two basic and complementary types of systems occur and each of these can be expressed in a number of ways. Tutorials develop within storybook reading, as do personal systems for exploring and playing with the activity.

Proposition Six: What children learn to do with storybook reading is become relative experts reflecting the nature of the activity.

The next chapter extends this analysis of activities to early writing.

Implications

For families, educators, and researchers

> Reading storybooks is important to school forms of literacy.

Families: Reading storybooks in ways which focus on the story, which build on the language of the book, and which make connections with children's experience and knowledge are important to school forms of literacy. These ways can be used without replacing or downgrading other ways of reading books which are important to the family.

> An important educational role is to foster storybook reading in collaborative and mutually supportive ways, and not undermine significant family practices.

Educators: An important educational role is to foster the development of storybook reading in families which has the features of collaborative participation focused on narrative meanings. However, this role has important criteria — to educate in collaborative and mutually supportive ways, and to educate in ways that do not undermine significant family practices.

Researchers: A priority for researchers is to develop further understandings of the social and cultural processes in different book reading activities. Research is needed regarding the features and conditions of collaborative and supportive educational programmes.

Further Reading

Meanings in storybook reading:

Duranti, A. & Ochs, E. (1986). 'Literacy instruction in a Samoan village'. In B. B. Schieffelin & P. Gilmore (eds.). *The Acquisition of Literacy: Ethnographic Perspectives.* Ablex, Norewood, NJ.

Gee, J. P. (1990). *Social Linguistics and Literacies: Ideology in Discourses.* The Falmer Press, London.

Heath, S. B. (1982). 'What no bedtime story means: Narrative skills at home and at school'. *Language in Society,* 11, 49–76.

Developmental outcomes of storybook reading:

Dickinson, D. K., De Temple, J. M., Hirschler, J. A. & Smith, M. W. (1992). 'Book reading with preschoolers: Coconstruction of text at home and at school'. *Early Childhood Research Quarterly,* 7, 323–346.

Sulzby, E. & Teale, W. (1991). 'Emergent literacy'. In P. D. Pearson, R. Barr, M. L. Kamil & P. Mosenthal (eds.). *Handbook of Reading Research, Vol 2.* Longman, New York.

Collaborative research models (in New Zealand):

Bishop, R. & Glynn, T. (1992). 'He kanohi kitea: Conducting and evaluating educational research'. *New Zealand Journal of Educational Studies,* 27, 2, 125–136.

Smith, L. (1991). 'Te Rapunga i te Ao Marama: Maori perspectives on research in education'. In J. R. Morss & T. J. Linzey (eds.). *The Politics of Human Learning: Human Development and Educational Research.* Longman Paul, Auckland.

End of chapter notes

1. Wolf and Heath (1992) discuss a related set of meanings. They argue that in acquiring knowledge children also acquire values about knowledge which are closely linked to the development of intentionality. Their comprehensive descriptions of children's literature and family literacy in a middle-class family include the claim that children learn to search for and decipher possible intentions for the characters in texts, that they can attribute intentions.

2. Ochs (1982) argues that the ways in which parents elaborate and expand upon their children's utterances reflect a belief in the young child as having intention. It is a belief that children intend to speak and intend to communicate. Moreover, parents believe it is appropriate for them to try to understand and adapt to the child's perspectives and needs (Ochs 1982). These characteristics are reiterated by Walkerdine and Lucy (1989) in their analysis of interactions between middle-class mothers and their 4-year-olds. They argue that spoken language provides a means of social (and textual) control, even for children.

3. The research is difficult to interpret because comparisons often have not controlled for differences between families in book-reading experience and types of texts used by families, and the type of text has been shown to influence interaction patterns (Pellegrini *et al.* 1990). Some researchers who have deliberately introduced narrative texts into the literacy practices of African American families in low socioeconomic groups, have noted the similarity between subsequent interactions and those in Anglo middle-class families (Heath & Branscombe 1986). Others, however, have reported that exchanges typical of collaborative participation tutorials were infrequent in those families who had been introduced to narrative texts (Pellegrini *et al.* 1990); and there is recent evidence that parental ideas about appropriate pedagogy and the functions of the activity, as well as beliefs about the researcher's expectations, are stronger determinants of the form that the activity will take than the sheer presence of a storybook (Gallimore and Goldenberg 1993; Renshaw 1992). I discuss these issues further in chapters 9 and 10.

4. The eight families identified themselves as Tongan ('Mo'ui Fakatonga') and were part of one church group. Their income levels placed them at level 4 and 5 on the New Zealand SES index of 6 levels. There were two to five children in each family, and living in the household as well as were adults other than parents, and nieces and nephews (see McNaughton 1994a).

5. Greenfield's (1984) study in Mexico of Zinacanteco apprentices learning to weave is a particularly good example. Errorless learning of fixed patterns was linked to the functions of weaving within the society, such as the preservation of a few culturally preferred patterns, and to the expensiveness of materials. Greenfield also describes the manner of interaction as a tutorial pattern using the general idea of a scaffold because of the presence of the general features of dynamic flexibility and shifting support.

6. Kessen (1991) has a word of warning for psychological research which compares cultural processes. His advice is to always assume that we will underestimate the extensiveness and subtlety of cultural diversity in the products and processes of socialisation. Cultural messages are multifaceted. Also, within any group important differences will exist between families which show multiple forms of socialisation at work (Valsiner 1994a). I have this in mind when contrasting meanings in different groups.

7. Metge (1984) points out that both in traditional and contemporary contexts the form and content of expertise are continually adapted to serve the needs of their time. 'The true puukenga does not simply reproduce what he or she knows factually, does not just sing a song or recite a whakapapa. He or she reviews all the similar situations he has witnessed and decides how to act in this one, which song to sing, which ancestor to begin with, by a rapid computing of all the variables involved'(p. 14). The account by the missionary Wade in Chapter 5 (p. 95) of the prodigious memories of Maori he met, goes on to include descriptions of how the same people would take learned phrases and play with their meanings. This expert use of 'highly figurative, and strangely elliptical language' (p. 129) obviously confused the missionary who was not an expert in this use of oral language.

8. This illustrates the close interweaving of cultural meanings in activities. It also illustrates the repetition of these messages across activity settings producing what Valsiner (1994a) claims is an essential condition in socialisation — high redundancy of cultural messages. The redundancy is necessary because of the many voices in a multicultural society. Given the active reconstruction of socialisation messages by the child, multiple messages are necessary. These are needed to provide rich and repeated information that encourages personal construction processes to take place. In the present case the Bible reading carried mutually reinforcing meanings about authority and about group cohesion, which parallels messages about group cohesion in other settings.

9. The argument parallels recent theoretical discussions about socialisation processes. A number of writers have discussed core features of effective scaffolding which may be universal to socialisation tutorials. They, too, assume the presence of underlying frameworks and local adaptations (Rogoff 1990; Lave & Wenger 1991).

10. Being able to understand and represent cultural meanings in research is difficult for persons outside the cultural group. The 'uncultured' researcher cannot know what is necessary to know. At the very least there is a requirement to develop familiarity with the forms of expertise one is studying in terms of its cultural meanings (see McNaughton 1994b). Being an insider, however, is not a guarantee that meanings will not be distorted and effective research processes are pursued either, as Valsiner (1994b) has pointed out. Being an insider, though, does create both the basis for access to valid forms of knowledge and processes of checking and clarification that are not immediately available to an outsider.

Chapter Seven

Early activity systems: writing

> **Focus**
>
> **Activity systems for writing**
> - As with Chapter 5 the descriptions in this chapter are based on Propositions Three, Four, Five, and Six.
> - A framework for describing activity systems for writing is presented.
> - The framework is based on descriptions of the participants' ideas, the interaction patterns, the tutorial types, and the materials used.
> - A set of activity systems for writing is described, together with the typical expertise which develops with the systems.

Chapter 5 described activity systems for reading and the development of expertise within them. This chapter provides a general description of writing activities. One of the activity systems is writing names so this chapter incorporates the descriptions from chapters 3 and 4 as well. The same four dimensions are used as were used for exploring reading: the participants' ideas and goals, their interaction patterns, the overall tutorial type, and the typical materials used. The concept of an activity system again is central to the description. An activity system is an activity with tutorial properties which give it developmental power.

Four Dimensions for Writing

1 Ideas and goals

Activities for writing, just as for reading, are purposeful. Learners and family members develop intentions within activities. Some of these activities share common goals with reading, such as learning to discriminate and name letters of the alphabet. But specific goals for writing include learning to write one's name for a variety of purposes, such as establishing identity, and being able to create personal accounts using written language. Ideas about writing, about learning, and about the nature of development are expressed and constructed in the activity by both the expert(s) and the learner.

> Ideas about writing, about learning, and about the nature of development are expressed and constructed in the activity by both the expert(s) and the learner.

2 Interaction patterns

Exchanges which become routinised are the building blocks of activity systems. In writing activities familiar ways of interacting occur which can be described in terms of form, content, participation, and instructional properties. The exchanges in writing activities share the three basic forms with reading activities of labelling, conversational or narrative, and performance routines.

> The exchanges in writing activities share the three basic forms with reading activities of labelling, narrative, and performance routines.

3 Tutorial types

Activity systems have particular configurations that give them a learning focus and developmental potential. Three basic configurations were described for reading — collaborative participation, directed performance, and item conveyancing (see pp. 73–74). These types also occur with writing.

> Three basic configurations in writing systems are collaborative participation, directed performance, and item conveyancing.

4 Materials

What children write upon and with is part of the identity of an activity. The tools a child uses help define purpose and thus constrain what is written and how it is written. As with the materials for reading, the definition of the material is partly given in its obvious properties (a blank sheet versus lined paper in a writing pad). But in many respects what is afforded by materials is given socially. Whether the piece of paper represents something upon which to sketch out a story, or something on which to write letters of the alphabet is dependent on intentions and the history of use of similar materials within tutorials.

Some tools for making marks are common to many activities, such as pens and pencils, chalk, and felt-tip pens. Other materials are adopted and adapted for the purposes of writing. Megan Goodridge, an educator from Barbados, describes children playing with whatever material might be at hand and using sand and mud to make their messages (Goodridge & McNaughton 1993). I recounted in Chapter 2 a mother's observation of her child making letters with water on the concrete path outside the home.

In general, materials can be grouped into several sorts. There are materials used for creating texts, such as stories. There are materials for producing item arrays, such as lists, and letters; the medium with which these items are made may be the same as that used for stories. Or they may be specially manufactured materials, such as magnetic letters which are placed on an unusual material such as a refrigerator door. Materials provide structuring for the activity. Some materials such as magnetic letters are very specific to purpose and their use provides considerable structure and focus for the activity. In this sense their properties provide a physical scaffold for the marking of the letter.

Exploring Activity Systems for Writing

Writing names

Tutorial systems

> Writing one's name has significance for the development of writing for other activities. This parallels the general significance of storybook reading.

Writing one's name is a very general writing activity. Its significance in the development of writing for other activities, including writing at school, parallels the significance of storybook reading. Naming appears in a number of specific activities which, because they have distinct goals and intentions associated with them, need to be recognised as separate. For example, children write their name (their signature) in the context of writing a letter to someone in order to identify themselves; they write their name when naming a painting so that they can express possession.

Nevertheless, the research evidence suggests that apart from their distinct purposes, when tutorial systems are involved they are configured in similar ways. The routines are therefore the same and the developmental outcomes, at least for what can be written, are the same. And the general intention to name in each of these activities is the same. Also, strong evidence suggests that learning to put one's name in writing is an early and perhaps basic activity in many families which generalises to these other acts of naming (Ferreiro & Teberosky 1982).

> There is strong evidence to suggest that learning to put one's name in writing is an early and perhaps basic activity in many families which generalises to other acts of naming.

The significance of this early naming is underscored in a recent study of 18 Maori, Pakeha, and Samoan families (Goodridge & McNaughton 1993). Over 800 writing products were collected representing writing over a six-month period from the time children in these families were 4 years 6 months old to the time they turned 5. Writing one's name was the major product for each of the groups of families and all the children produced examples of writing their name at some level of expertise. It seemed that this activity was particularly strongly represented in the Pakeha families, as nearly 20 examples on average per family were collected from five visits. But this may also have been a consequence of the practice in the families of collecting the examples of attempts at writing a name.

Children write their names on many materials. The descriptions of early writing in families show that in most families materials and tools are largely irrelevant to the performance. One can teach and learn to write a name with almost anything that can be marked and comes to hand.

Some materials, however, are stronger 'setting events' for adding one's name. Letters written to others are an obvious activity in which there is a required slot for writing one's name. Similarly, drawings and other creations of children often are named (signed). This may be a product of early-childhood education experiences where distinguishing individual children's productions within a group is a necessity. It may also reflect a wider value of needing to own one's production.

Many parents in Western industrialised countries articulate a goal for children to write their name, and attach an expected age for this to be achieved. This is often, but not exclusively, associated with white middle-class families, and families differ in whether they expect children to achieve this before going to school (McMillan 1984; McNaughton & Ka'ai 1990; Phillips 1986). In the study which produced the 800 products, each family, irrespective of cultural group, believed that it was important for a child to learn to write their name before going to school. Similarly, in the SOL study (see Chapter 2) all parents accorded it the same status as the activity of book reading in their family practices. By 5 years of age all but one of the children who could write a word wrote their name or part of it (in the form of the first letter). In other words, the probability of the first word able to be written being a name was very high.

At the level of underlying cultural messages, naming can be linked to beliefs about developing a personal identity, being an independent agent, and being able to act on one's own (see Czerniewska 1992 & Wertsch 1991). This need may underlie the widespread practice of adding a name to a letter. Families in each of the three communities Heath (1983) described wrote letters, and it appears that children added their names. One child, Mel, when asked to write his name at 4 years produced lines and marks representing the date, a salutation, the body of the letter, and a closing. He wrote his name at the bottom of the page. There are no observations of how this or other writing of names was accomplished, although it is clear from Heath's comments that it occurred as a collaborative activity. It is hard to see how the right way to write a name could be discovered.

> At the level of underlying cultural messages, naming can be linked to beliefs about developing a personal identity, being an independent agent, and being able to act on one's own.

At a more prosaic level, the initial purpose for the tutor is to convey the message that one's name can be represented physically. Over time the intentions of both learner and tutor come to focus on producing the right way to render the name.

What little systematic information does exist suggests varieties of demonstration and labelling routines are employed. The SOL study contains instances of the modelling of names in ways very similar to the examples given for Harry in Chapter 4. All of these instances and the model plus imitation are formally equivalent to the demonstration routine used in Reading for Performance (see Chapter 5, p. 92).

The sample on page 128 is an example of writing from one of the families. In her diary, Jeremy's mother wrote how she had initiated one activity but, following his negotiation, ended up with the naming activity. Her diary notes record: *I write the A B C for him to copy but he said I [he] write something else.* The 'something else' was a series of sketches of people and objects. Jeremy's mother added his name at the top of the page and Jeremy copied twice. Given his attempts, she also provided a partial model underneath, in the form of an /e/ (in the right orientation) and an /m/ (which had been missed out). Jeremy then added the other letters.

> Models accompanied by imitation are well suited to trying to teach someone to write a name correctly. A letter may be formed in many ways, but the standard form must be acquired.

Models accompanied by imitation are well suited to the job of trying to teach someone, especially a 'little' someone, how to write a name correctly. While a letter may be formed in many ways, the standard form must be acquired. There are limits on ways to write a name so that most others will recognise it.

Information was not collected in the SOL study on the detailed moment by moment interactions around the writing of names. Interestingly, doing this poses a far greater challenge to researchers' time and the ingenuity of their measures

Sample 7.1 The copying of his name by Jeremy, aged 4 years 6 months

> The general function of interactions is to facilitate the segmentation of the whole name, the identification of its elements, and the production of elements in the right sequence. Alphabetic and phonemic elements are identified, and algorithms are deployed to achieve this.

than examining storybook reading. Writing does not occur as regularly as some reading activities such as bedtime stories. The intention to write is less often linked to regular events, following on other events, such as the reception of a letter or the need to sign a birthday card. Moreover, a strong theoretical tradition stressing the active self-discovery by children has provided a rationale for researchers to not search for tutorial properties in interactions involving the writing of names, anyway (see Chapter 1).

In collected samples of the writing in the SOL study, however, there was evidence of many modelling opportunities created by the tutor (caregiver). Similarly, evidence from our testing showed a strong tendency for a single letter to stand for a name, suggesting the presence of graduated and embedded interactions.

When interaction patterns have been described in this activity it is apparent that within and complementing the demonstration routines are a number of other interactions which highlight features of the task. These interactions are a mixture of demonstration and display routines. Their presence, and especially the degree to which explanation and elaboration are provided alongside of them, may be a source of differences in how families carry out the activity. Their general function is to facilitate the segmentation of the whole name, the identification of its elements, and the production of elements in the right sequence. To achieve this, alphabetic elements (e.g. H-A-R-R-Y) and phonemic elements (e.g. Huh) are identified and algorithms for the production of letters and sequences are deployed.

The presence of any of these sub-routines adds value to the configuration of the system. The value is in the form of making connections between sounds and their graphic representations on the one hand and motor control for producing letters on the other. These routines have the characteristics of demonstration and labelling. But they are substantially modified by uses of language to provide reasons and to negotiate when and how they are used.

Alphabetic routines in their various forms provide an oral spelling commentary together with the model. The most demonstrative and informative form of an alphabetic routine involves exaggeration of the letter names, with clear pausing between letters. The accompanying commentary is closely matched to the formation of the letters in the model. In its simplest form, at the mid point in the system's development towards full independence by the learner, the written letters may be written, and their names may be spoken, quickly. Or before writing, the expert may cue the letter names (*'Remember H-A-R-R-Y'*). Over time, the system reduces the commentary and the model until neither are needed.

Phonemic routines add sounds to the model, for example, *'/H/ ("Huh") for Harry'*. Again, in such a routine's most demonstrative and informative form the sounds are exaggerated, the pauses between the sounds are lengthened, and the visual association between grapheme and phoneme controlled. In the case of the many irregular and difficult associations between letters and sounds, the commentary and highlighting might focus on the simplest chunks to hear, including consonant clusters in the onset of the word and syllables in the ending rhyme e.g. *'Chr – i – s – tine'*.

Letter-production algorithms provide a simple rule for producing a letter. They come in two forms, one in which the learner is talked through (e.g. *'straight down and across for /H/'*), and one in which the learner is physically guided through the production.

The tutorial configuration in the activity of letter writing, especially in the early stages, is one of directed performance. But as with the use of particular sorts of books in reading, materials can be used which provide physical structures for the developing expertise. Heath (1983) described the children in the white middle-class families with whom she worked going through several general stages in their experiences of print. As the beginning of school approached children entered a stage which was a preparation for school. Heath records how:

> They are given workbooks and encouraged to write their names ... These adult-supervised experiences reinforce repeatedly that the written word can be taken apart into small pieces and one item linked to another by certain rules. Children gain manipulative experience in the linear sequential nature of books. Parents tell them to follow simple rules: stay in the lines, write answers on the lines, begin at the beginning, match the cutouts of letters and shapes to corresponding diagrams. (p. 228)

Four phases are discernible in the tutorial. The first is the joint establishment of an intention to write a name. The second is the provision of the full configuration with embedded routines. The third is the transition to minimal forms of the tutorial (for example, when the only input for a parent is the cue: *'Remember how you write your name?'*). The final phase is the development of independence in being able to write a name by oneself.

Together with their embedded routines, writing tutorials provide a complex supportive structure. This complexity and its flexibility within an episode and over time is shown in the transcripts on pages 130 and 132. They record a 4-year-old girl and her mother at two different times, four weeks apart (Kempton 1994). The activity system had developed further than performance from simple models and, therefore, is different from what I described for Harry in Chapter 3. The routines were a mixture of display and indirect performance routines. Nevertheless, there are some obvious similarities that show how the earlier form develops into one that is described in the dialogue below.[1]

The most obvious feature of this episode was the general presence of sensitive and contingent adjustment to suit the learner's needs and goals. Four levels of adjustment were discerned in the tutorial representing a different focus of support for the performance. The levels ranged from a focus on items (level 4: Letter production) to a focus on production of the name as a whole (level 1: Name production).

Sample dialogue 7.1
Tutorial for writing a name. Performance and display routines

Episode 1

Name production (level 1)
MOTHER	Do you want to write your name?	(QUERY)
CHILD	[writes unconnected letters over blackboard]	(DISPLAY)
MOTHER	No. Those are letters.	(FEEDBACK & LABEL)

contingent adjustment

Letter production (level 4)
MOTHER	What are those letters? [pointing]	(DEMONSTRATION)
	Can you draw an /M/?	(QUERY)
CHILD	I drew a /M/.	(DISPLAY)
MOTHER	An /M/ .	(FEEDBACK & LABEL)
	Can you draw an /A/?	(QUERY)
	Can you draw a /Y/?	(QUery)

contingent adjustment

Spelling (level 3)
| MOTHER | How do you spell your name? | (QUERY) |
| CHILD | A ... M ... Y | (DISPLAY) |

contingent adjustment

Sequencing of letters (level 2)
MOTHER	Can you write Amy?	(QUERY)
CHILD	[writes /A/ and /M/]	(DISPLAY)
MOTHER	How about a /Y/?	(FEEDBACK & QUERY)

In the first episode the child was writing on a blackboard. The mother orientated her to the task with a query which functioned as an indirect request to perform what had, in earlier sessions together, been modelled. She was asked to write her name. This was categorised as a focus at a first level, that of **Name production**. However, the child wrote unconnected letters, (referred to in the name-production part of the transcript at level 1). This resulted in the mother adjusting the level at which they focused their efforts to the fourth level, **Letter production**, at which she was to identify letters. A series of interactions followed which were forms of Display routines (*'What are those letters?'*) and also indirect models for performance (*'Can you draw an /A/? Can you draw a /Y/?'*).

After this the mother adjusted the focus again. She did this by using a query as an indirect request to perform the alphabetic sequence which provided an

Amy's initial production for writing her name in Episode 1 (in Sample dialogue 7.1)

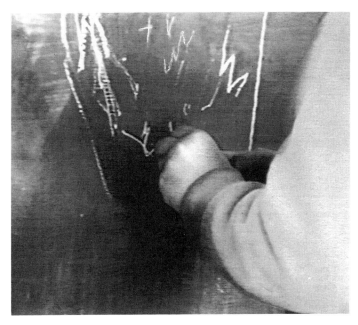

oral model for the sounds in her name (level 3: **Spelling**). This lead to a further adjustment to a focus (at level 2) on the **Sequencing of letters**. Having rehearsed the sequence verbally the mother requested the child to write the sequence using the shared (agreed upon) spelling as a guide.

The flexible and dynamic support in the first session produced the recognisable letters of Amy's name. An approximate sequence was achieved but the /A/ was upside down and the letters were spaced around the blackboard. Most of the tutorial focus, in terms of the percentage of interactions spent at each of the levels, had been at level 4 and level 2: letter production and sequencing). This information is summarised in Table 7.1.

Table 7.1 A mother and 4-year-old writing a name: the percentage of interactions at each level in two episodes four weeks apart

Episode		One	Two
Levels			
1	Name production	6.8%	39.0%
2	Sequencing of letters	47.7%	30.5%
3	Spelling	9.1%	30.5%
4	Letter production	36.4%	0%

(Source: Kempton 1994)

After four weeks the overall focus of the tutorial had shifted. This paralleled developments in Amy's expertise. Part of the transcript of an episode a month later is shown on page 132. Amy's first attempt at writing her name on a picture (not shown) resulted in only an /A/ being written because it was squashed on the left of the picture. This first part of the episode is not shown in the transcript. After this first attempt the mother provided a model which Amy copied.

Sample dialogue 7.2
Tutorial for writing a name

Episode 2 (4 weeks later)

MOTHER Can you write it in that order? Start over here [pointing to the bottom left of the page] so you've got plenty of room over there down the bottom. Do the /A/ and the /M/ and the /Y/ [pointing to the space on the page where each letter should go]. Can you do that? If you start over there then you'll have room and then you can give it to J … [a friend]
CHILD [writes /A/]
MOTHER Good. What goes next?
CHILD /M/.
MOTHER Good. Put the /M/ right beside the /A/. Then it'll make a proper word.

Amy's second attempt (below) shows how critical features which were highlighted in the task were to do with sequence within a defined space of a piece of paper (*'Start over here'*). Amy and her mother concentrated their energies on getting the sequence right and her name produced. The resulting productions at this time almost have the standard orientation. In terms of the interactions in the whole episode they had split their time between spelling (level 3), sequencing (level 2) and producing the name (level 1).

Sample 7.2 Amy's production of her name in Episode 2

This dyad was part of a larger study of both Pakeha and Maori families in which the parents had a range of educational levels and occupations (McNaughton, Kempton, & Turoa 1994). The similarities and differences between these families are important because they represent ways in which the pedagogy and goals of the activity express cultural messages. One major difference between the families was apparent in the roles of other family members. Whether or not sessions were typically dyadic or multiparty, paralleled those differences found in the studies of storybook reading. In other words, writing in Maori families, as with storybook reading, tended to be multipartied, reflecting the messages about the role of the individual and expertise discussed in Chapter 6. Writing in Pakeha families, however, was a dyadic activity.

Differences were apparent also in the role of personal initiative and the structuring of the activity. In some Maori families the tutorials had a more open structure than in most Pakeha families, and the exchanges were not as focused on particular learning at each of the levels. Some Pakeha families tended to reflect more on the reasons for performance and negotiate these, as in the discussion about spatial orientation in the example of Amy's second episode (p. 132). In some Maori families the negotiations between child and tutors tended to be focused on the degree of demonstration which might be required. But, apart from the participation structure families in the study were more similar than different.

Personal systems and the development of expertise

What develops within this writing activity system? The specific outcomes are easily recognisable. Children come to be able to write their names. This general expertise is embedded in several different sorts of activity, including adding names to picture and letters, as well as writing it on request for others to see. While this latter example could be considered to be writing without a context, it is, in fact, a different sort of context, but also with purposes, rules, and expectations.

Until recently this competence was assumed to be constructed by the child from ambient stimulation (Ferreiro & Teberosky 1982). The prevailing constructivist theorising (p. 12), together with the sheer difficulty of capturing fleeting activities embedded in everyday life, directed research attention to looking for what children could do by themselves, particularly displaying whether they could write their own name. The mounting evidence, however, supports a co-constructivist view, which sees this expertise developing within tutorial activities as well through personal constructions from ambient information.[2]

Learners also develop generalised strategies for writing names and other words. As with the developmental effects of learning to label, a significant knowledge base develops, which is the understanding that a name can be represented in written symbols. Depending on the presence and characteristics of the embedded routines, the learner also comes to know a 'big' thing, that there are relationships between sounds and letters. The start to this awareness comes from knowledge about specific relationships (*'There's an /H/ for Harry!'*). In some forms this knowedge can contribute to the development of **phonemic awareness**, an awareness that one's language can be broken into sound units (see Goswami & Bryant 1990). Recent research links this knowledge with effective learning in beginning reading instruction (Sulzby & Teale 1991).

The specific spelling expertise and more generalised knowledge about sounds and letters is used strategically by the child to personally construct further expertise in writing. These outcomes for performance routines were depicted in general in Chapter 4 (p. 73). Using this outline, the more specific outcomes for this activity system are shown in Figure 7.1 (p. 134).

> Evidence supports the view that writing one's own name develops within tutorial activities as well as through personal construction.
> Learners also develop generalised strategies for writing names and other words.

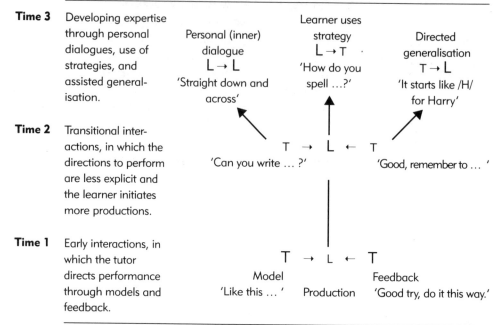

Figure 7.1 Developing writing expertise, based on learning to write one's name in tutorials of directed performance — with tutor (T) and child learner (L)

Three sorts of outcome are shown. One is the strategic control by the child to gain needed information (for example, the child can ask, *'How do you spell ... ? How does ... go?'*). Another is a generalised capability of the activity system. The tutor and child can utilise the specific knowledge further. For example, the spelling routine is incorporated into a strategy that the child can then use to get models for further words. The tutor may use generalisation to a new word saying, *'It starts like /H/ for Harry'*. There is also the use of routines, strategies, and algorithms for personal scaffolding, as in the personal cuing of *'straight down and across'*.

Schickedanz (1984) worked with families of six preschool girls who were reported to have an intense interest in literacy (ages between 3 years 6 months and 5 years). The mothers of these girls kept diaries. The diary transcripts provided by Schickedanz have many instances of the children asking how things, including names, are spelled. The following is one example:

> Then at the dinner table last night, she, I don't remember how it came up, but, she wanted to know her last name, how to spell it. So Don said, 'Well, how do you spell it?' She started out and it was wrong. Don said no and they spelled it out. He broke it down into its syllables. They did it out loud about four times. Then he gave her some paper and she wrote it out. She got it. She'd been playing around with it for a long time. It finally clicked last night. (1984, p. 17).

Even more specific knowledge and performance strategies may emerge in the developing system. The following account is from another of the families which supplied diaries for Schickedanz. The diary record shows the modification of knowledge about the possible ways to write a name. The developing system was triggered by an ambient event (the mother writing a letter). It shows the generalised power of the demonstration configuration and its flexibility. In the face of difficulty an attempt to structure the copying of a model was modified to one in which the child could trace.

> She had seen me writing a letter to her grandmother. She was interested in the script. She wanted to know what I was writing and I read her some of the sentences. Once I said her name and she wanted to know where it was. I showed her, but she didn't recognise it. She wasn't used to the script. So then she wanted to write it that way, so I did it on a piece of paper. But she couldn't do it, so I suggested that she put another piece of paper on top, to copy it. I was trying to finish my letter. Well, she tried to copy, but had no idea of the order in which to copy. Then she realised that she could write her name three ways now: manuscript upper case [capitals], manuscript lower case [small letters], and script, or what she calls the 'hard way'. So she went off and said she had a project, and would surprise me. This pleased me because I was trying to finish my letter. She went to the table and didn't want me to watch her. She took out a pad of paper and then came back and said she had done her name three times at the top of the paper. The first one was capitals. The second was small letters. The third she couldn't do, so she came over and asked me how to write her name in script. (1984, p. 14).

There is another very significant personal outcome of this name-writing activity. It is the deployment of the knowledge from this system and the expertise gained in other activities, such as forming letters in 'invented spelling'. Invented spelling is reported in many studies of children's development of writing skills before school. I describe this at the end of this chapter. It represents a generalised system which develops out of the more specific activities discussed here.

One further outcome can be identified in the development of writing at school. Being able to write one's name on entry to school is related to later achievement in writing at school. This relationship is not automatic. It takes place through processes similar to those that associate storybook reading with achievement in reading at school.[3]

> Being able to write one's name on entry to school is related to later achievement in writing at school.

Forming letters

Tutorial systems

Several studies of families in New Zealand have found them to have strong beliefs about the need to write letters of the alphabet before school. These are related to general views about the nature of conventional cultural tools. The major rationale articulated by family members is one that is pressing. It is that this is a way of gaining early reading and writing skills in association with learning letter sounds which will be needed for success at school. One mother stated this clearly: *'I want her* [her 4-year-old] *to know the alphabet so that she would have a good start at school'* (Goodridge & McNaughton 1993). I reported in Chapter 2 on how this was a highly visible tutorial activity for the parents involved in the SOL study.

> A belief articulated by some families is that writing letters is a way of gaining early reading and writing skills in association with learning letter sounds.

> Goals are dynamic and parents use available and salient sources of information to reconstruct continually their beliefs about goals.

These are not the only goals one might have for learning to write the letters of the alphabet. Parents can also articulate longer-term goals, such as noting that being able to write letters is a part of being able to write other things. Goals are dynamic and parents use all available and salient sources of information to reconstruct continually their beliefs about goals. Faced with particularly salient information parents modify their beliefs. For example, a Samoan mother reported how she modified her beliefs about alphabet writing in the light of her son's attempts and the input from other family members (Goodridge & McNaughton 1993). She had been teaching her child to write lower-case letters for his name which we had observed on the first visit. On a later occasion, however, she showed us an example of upper-case writing. She told us that the child's aunt had introduced him to capital letters by writing his name for him. The child traced them and then copied them. The mother reported he successfully wrote his name after that noting that *'The upper-case letters are easier for him'* (Goodridge & McNaughton 1993).

The standard exchange at the beginning of this letter-forming activity is a demonstration routine. When parents have reported on their teaching of letters, and when they have been observed, the common pattern has been a model plus tracing or copying (McNaughton & Ka'ai, 1990; Goodridge & McNaughton 1993). In addition, instances have been observed of more direct structuring of the model to include a model for how to produce the letter. In Chapter 3 this was identified as a **production algorithm** (p. 45).

In keeping with the properties of directed performance, the tutorial support shifts over time. In some instances this means careful reduction of the cues in the algorithm. In other instances it involves cues to remember joint productions. The recovery of the language that has been shared previously acts as a rehearsal prompt for the child without direct support.

Personal systems and the development of expertise

Clearly children come to be able to write letters. These may be embedded in their names initially, but under their own practice and play with letters (see Chapter 3) they come to be able to remove letters from one dominant context and generalise the use of those letters to other contexts. This expertise contributes to the general significance of learning the letters of the alphabet. Beyond this, the expertise entails the knowledge I described earlier associated with coming to write one's name (e.g. pp. 133–135)

> Children become able to write letters, initially their names, but under their own practice and play, they learn to generalise the use of letters in one dominant context to other contexts.

Writing narratives, descriptions, and expositions

Tutorial systems

Significant amounts of writing also take place that are narratives, or more formal descriptions and expositions. Families give children opportunities to comment on and describe their experiences. These may relate to personal events, significant persons, animals, places, both real and imagined.

A particularly telling example of this sort of text comes from a Samoan child (Goodridge & McNaughton 1993). It illustrates the direct effects of cultural values on what this boy chose to write about. Just before going to school he wrote, *I said to Sologa, you're a great Dad.* An older sibling collaborated with the child, Lilo, and actually wrote the text. Lilo copied the text underneath. Lilo constructed these orally based texts for his father to share with others at family gatherings and church. Lilo often constructed texts in English, particularly with his older siblings who attended primary school. The content of the texts was based on readings from the Bible and was about the goodness of people and goodwill toward them. The text on which he and his brother collaborated produced written language which had similarities with that in biblical texts.

This observation illustrates the presence of the demonstration routine (in the form of *'Copy what I write'*). But it is clear that the overarching form of the tutorial is one of collaborative participation. Conversational (or narrative) exchanges occur in the production of narrative, expository, and descriptive texts. These routines facilitate the co-construction of language suitable for writing down, the language base for the written product.

There are, however, few research examples of these collaborative products. Some of the reasons for this were discussed earlier. Times of writing are less fixed than times for reading books to children, so it is more difficult for researchers to be present and capture these interactions. Megan Goodridge (Goodridge & McNaughton 1993) describes how a Maori child (4 years 7 months) collaborated with her grandmother to produce a sketched map with dimensions approximately one by one and a half metres. After the significant landmarks such as the corner shop were discussed and put in place, the grandmother wrote on the map, *This is the way to Rumatiki's house* (below left). Rumatiki turned the page and traced parts of the grandmother's writing, which had been imprinted on the newsprint paper. Rumatiki soon engaged her Grandmother in the production of another map of similar dimensions. This time she drew her house on the map by herself and wrote her name beside it (below right).

The activity of making sketch maps in this family was significant, culturally and socially. The grandmother said that almost every Saturday the family went

> The guidance provided for narrative-writing is that of collaborative participation. This enables the production of narrative, expository, and descriptive texts.

Rumatiki's map: an example of collaborative participation in the production of a complex piece of descriptive writing and drawing (child aged 4 years 7 months). Left: the initial map; right: the child's annotated map

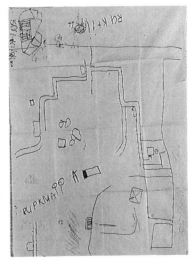

to garage sales and they collaborated regularly with others in the family or the community to produce sketch maps to reach their destination. The grandmother often accompanied Rumatiki for these walks between her house and the child's. The grandmother would point out the significance of landmarks along the way and, in the process, Rumatiki learned to do the same.

The cultural threads in this activity include messages about relationships in the family and about knowledge of one's locality. The intergenerational relationship itself is very important in Maori families, and knowing about the land and what landmarks signify is also a significant cultural marker.

Tutorials for narratives start with collaborative participation with performance routines (for example, on how to write letters) embedded in the collaboration. The language base is essential to the production of narrative and expository writings. Initially family members may write the text and may even identify much of the oral base. Gradually through the collaboration and through imitation and personal constructions children come to write the text independently, without an oral base.

> Through collaborations and personal construction children come to write texts independently, without an oral base.

A familiar sequence early in the establishment of the tutorial is for a family member to ask a question about the child's intentions in a picture: *'What is that? What have you drawn?'* In its simplest form this leads to an account or a description which the more expert person accepts as a dictated narrative or description and writes it. Collaborative participation is present in the initial dialogue followed by an interesting reversal of roles as the more expert person accepts the model provided by the child. It is at this point that differences between families may be significant, particularly in the extent to which the other person shapes and elaborates the child's account. Later on the initial part of eliciting an account is replaced by the child initiating an account.

Personal systems and the development of expertise

Development in this narrative/exposition writing activity is closely related to the developments from reading for narrative and exposition (see Chapter 6). The two activities are likely to provide complementary settings for the development of expertise with narrative and expositional forms of language.[4]

In addition, some more general outcomes have been assumed to develop. Together with other activities which are focused on extended sequences of text (such as making messages and reading for narrative), writing is seen as contributing to ways of thinking, primarily in the development of language as a tool for reflection and analysis (Olson 1991).[5] The more immediate outcomes are that children come to be able to preformulate, that is, to structure ahead of time, their experiences and their imagination in ways that can be written down. Among other things this means that they develop ways of modifying their oral language so that it is compatible with written conventions.

An example of the development of expertise and strategic use of the activity occurred in Megan Goodridge's descriptions. She describes a Pakeha child who commented on a significant domestic event. The family was renovating and the mother was busily involved in rebuilding the kitchen. She constantly was giving the preschooler and two toddlers strict warnings to stay out. The preschooler drew a picture of a lady with a crooked smile, and then dictated her experience to another person in another setting. She brought home the picture and writing and gave it to her mother. It said, *The monster was walking around the kitchen.* This was the preschooler's interpretation of the situation. She had depersonalised the situation, using figurative literate language to personify her mother.

The mother planned to frame the writing and mount it in the kitchen, an act that illustrates the values of caregivers in this activity to respond semantically to their children's language. In this process they recognise their children as intending to communicate and these personal intentions are important to recognise.

Making messages

Tutorial systems

A significant activity reported in many studies of family literacy is making messages. In the first instance these are often formulaic endings or notifications in letters and cards, but they include 'Thank you' notes, and 'love from' messages accompanying birthday cards and letters. Their formulaic nature means that the **frame** (simple sequence) can be practised over and over again, and each time is an instance of meaningful communication.

> The formulaic nature of written messages means that the frame can be practised repeatedly, each time being a meaningful communication.

The goals of the participants develop within the activity. Children come to know what cards and letters are used for within their families. The intention to thank, appreciate, and express love are important family goals. The research noted earlier by Schickedanz (1984) includes a report of a mother's goals for, and the development of this activity of making messages:

> I had written a whole lot of names yesterday on a piece of paper, kids in the class who she wanted to give Valentines to, and to Katherine and Stephanie, a bunch of people. So I told her to pick a name, and then write *'From C ... '*. At first it didn't go right. She wrote three names on one card. So I said, *'No, that's wrong. You put one name on each card and then put it in the envelope'*. I sort of stayed around and she was doing it right, that part. There were some long names. She was using big letters, writing big, and she ran out of room on the card to write them. So she stopped. I explained that she could abbreviate, but that normally you don't write half a name. (1984, p. 16).

Such frames (*From C ...*) provide a very important scaffold for this practice. They give a fixed structural basis for the emergent writer to practise and deploy their writing skills. The basic tutorial exchange is one of demonstration. But again, when complex messages are being produced, these are embedded in tutorials which involve collaborative participation.

> Frames provide an important scaffold for making messages, providing a fixed basis for the emergent writer to practise and deploy their writing skills.

Personal systems and the development of expertise

The tutorial configuration begins in the same way as the naming activity (p. 126) but increasingly becomes collaborative participation around co-constructed messages. As expertise in spelling develops, personal constructions take over. The same mother who supplied the diary record above went on to record another event:

Anyway, Sunday morning, early. She made me an Easter surprise and she glued little notes to an easter bunny. I'm not really awake yet and this thing's right under my nose. She says, *'Here it says "Easter".'* Little pieces of paper glued, taped, but it doesn't say all easter, just EST. I thought, where did she ... We don't have any easter cards around that she could have seen. So she said *'EST. Do you hear it?'* She thought this was it, it's only the a that's out. Then she said that she started out with her drawing and that she didn't have room for more writing. The 'e' and 'r' at the end she didn't get. The 'a' in the first part she wouldn't have gotten in any case. She would think EAST is EST. I was surprised, but she said, *'That's the way you spell it.'* I wrote it down for her. *'That's the way you really spell it.'* But she wasn't very interested. (Schickedanz 1984, p. 18).

This developing system has developmental outcomes similar to those in the developing system for writing names. Indeed each of these systems described above supports and complements the other in the development of personal expertise in producing writing. Coming to control frames enables the child to gain more control over letters and the strategies associated with coming to form words. Phonemic awareness is heightened by the degree to which letter–sound relationships are revealed or highlighted in tutorials. In this way, alphabetic and phonemic routines embedded in the provision of a model for the frame will contribute to that awareness. Nevertheless, there comes a point when the child has 'cracked the code' and uses all opportunities, irrespective of the presence of embedded routines, to try out and extend their knowledge. And this leads to generalised and generalisable expertise which has so impressed and fascinated researchers. It has been called invented spelling.

> In controlling frames the child gains control over letters and the strategies associated with forming words. Phonemic awareness is heightened by the degree to which letter–sound relationships are highlighted in tutorials.

Invented spelling

The Easter example above provides an illustration of an important developmental outcome for families. Each of the specific activity systems for writing feeds into a general expertise which is under the control of the learner. This expertise involves the development of a spelling system for producing writing in contexts of use. Under some conditions, such as when the inventions are valued in the family, this system can take on a very personal form, which is independently devised.

It is interesting to note that in writing activities, models for words and names are not typically given as approximations. For example, in my interactions with Harry described in chapters 3 and 4 I did not find myself teaching him to write HRY as an approximation. And I did not provide him a direct model for his attempt at writing seesaw (as SS). Yet a number of children develop personal systems which use invented spellings in the form of approximations for written sounds. Bissex (1980) and others have described in case studies how compelling and powerful this developing expertise is for children. Being able to create writing which has personal and interpersonal functions can be very reinforcing, for both children and their parents.[6]

> Each writing activity feeds into the learner's general expertise, which involves the development of a spelling system that may be a personally devised form.

There are important conditions for the development of personal systems of invented spelling. The case studies suggest these include intense experience with a wide variety of writing activities within which learning and development systems form (see also Wolf & Heath 1992).

Tutorial systems and personal systems merge into one another in the developing system of invented spelling. The learner's developing expertise enables them to try to write by themselves using their incipient skills. The intention to write by oneself, with attendant risks of being wrong, develops from the specific activity systems. The structure provided by those systems which produce emergent expertise creates the foundation. When conceived as part of a developing system the role of the family members becomes one of audience and responding to the meanings of what is written. They also become a source of contingent feedback of correct models and rules. As in other personal systems, ambient print provides a wealth of information about rules and models.

> The intention to write by oneself develops from specific activity systems. As part of a developing system, family members become an audience, and a source of response and contingent feedback.

Summary

Activities for writing can be identified. Like activity systems for reading, they have structural constituents and, under appropriate circumstances, effective tutorials develop. Writing one's name provides a firm basis for developing general writing expertise. The degree to which more generalised spelling develops is dependent on the range and focus of other writing activities which children experience. The propositions on which the two previous chapters were based have also guided this chapter.

Proposition Three: Literacy practices are expressed in specific **writing** activities which have identifiable properties. These include goals, rules for participation, and ways of carrying out the activity.

Proposition Four: Learning and development systems take form within **writing** activities as a product of the child's actions and the actions of significant others.

Proposition Five: Two basic and complementary types of system occur and each can be expressed in a number of ways.

Proposition Six: What children learn to do with written language is become relative experts within particular **writing** activities.

> Writing one's name provides a firm basis for developing general writing expertise. The degree to which more generalised spelling develops depends on the range and focus of other writing activities which children experience.

Implications

For families, educators, and researchers

Families: The same implications as those noted at the ends of chapters 3 and 4 also apply here. Families can recognise the developmental significance of children's experimentation and play with writing. They can guide writing both indirectly and directly through ambient and joint activities, and through providing opportunities for personal activities.

Educators: Educators potentially have a significant educative role in families' construction of ideas about writing and learning and development. Educators need to be aware of different families' needs for access to and participation in both intended and unintended educational knowledge and expertise.

> Families can guide writing both indirectly and directly by providing for experiences of ambient, joint, and personal activities.

> Models of collaborative educational programmes are needed which foster diversity without undermining significant cultural practices.

Researchers: Much less is known about the co-construction of expertise in writing than is known about reading. More research is needed on social and cultural processes which plots forms and sequences of expertise. The implication noted at the end of the last chapter can be reiterated here. Models of collaborative educational programmes are needed which foster diversity without undermining significant cultural practices in writing.

Further Reading

Writing and cognition:
Olson, D. R. & Torrance, N. (eds.). (1991). *Literacy and Orality*. Cambridge University Press, Cambridge.

Early writing processes and early education:
Czerniewska, P. (1992). *Learning about Writing*. Blackwell, Oxford.
Dyson, A. H. & Freedman, S. W. (1991). 'Writing'. In J. Flood, J. M. Jensen., D. Lapp. & J. R. Squire (eds.). *Handbook of Research on Teaching the English Language Arts*. McMillan, New York.

End of chapter notes

1. In this study involving seven Pakeha and 10 Maori families the families agreed to carry out writing sessions at particular times so that they could be observed. It could be argued that these episodes therefore were unusual. But the methodology attempted to minimise this by using activities that the parents reported were familiar and, at the time of observation, appropriate to engage in. Parental reports suggested the resulting sessions were minimally different from what typically took place.

2. Unlike reading activities there are few systematic analyses or experimental attempts which would show that the tutorials described here are instrumental in producing writing. But the exchanges and their effective outcomes, as well as the general correlational associations provide strong support for the conclusion that these tutorials are functional in the development of the expertise. In our research we have found substantial shifts in the expertise of 17 4-year-olds within tutorials over four-week intervals. Also, patterns of well tuned and less well-tuned tutorials were associated with more or less advanced forms of expertise (Kempton 1994).

3. One-quarter of the 340 English children in a study by Blatchford (1991) could write all of their name correctly on entry to school. The degree of expertise at this time was significantly related to later measures of transcription, grammaticality, and richness of vocabulary at the age of 7 years. Half of the parents had said that they taught their child to write their names and other words before they started school, and only 9 per cent did not, or only rarely, taught writing. This self-report, together with the measure of child's preschool writing, was what was significantly related to the measures of writing at 7 years, and not other related measurements, such as experience with books, and concepts about print. This specificity of effects on school progress of direct help with writing early on is suggested also in research in the United States by Snow *et al.* (1991).

4. Pappas (1993) has recently described how young children develop schemas for exposition as well as narratives. Her research effectively challenges an assumption that narratives are somehow prior or more basic to early literacy. We have found considerable differences between families in the degree to which narratives or expositional forms are emphasised. But many families in which writing is produced collaborate in descriptions and expositions and in one study we have found expositions, being produced by 4-year-olds at least as frequently as narrative writing (Goodridge & McNaughton 1993).

5. Olson's (1991) view is that ' ... writing takes language for its object and just as language is a device for "fixing" the world in such a way as to make it an object of reflection, so writing "fixes" language in such a way as to make it an object of reflection' (p. 266). More generally, he argues that writing turns language into an object of thought and analysis and increases awareness of relationships between representations of the world and language. The activity analysis suggested here argues that generalised expertise is closely related to usage. Given the widespread use in activities of written language for reflection and analysis, widespread forms of this mode of thinking about representations of the world and language would be expected (see also Scribner and Cole 1981).

6. Invented spelling is an expression of a basic cognitive endowment (see Chapter 3, p. 46). It is a strategic adaptation to the situation (problem) of writing. In this sense invented spelling involves processes similar to those in children's productive over-generalisations in oral language and strategic mixing of codes in the development of bilingualism (Lindfors 1987).

Part Four

Relationships

and

Transitions

Chapter Eight

What develops?

> **Focus**
>
> **The development of expertise during children's preschool years at home and in other socialisation settings**
> - Descriptions of the development of early expertise are available. These are bound in several ways: by the measurement procedures which have been used, and by the place, time, and manner of testing.
> - The descriptions show, however, that children become relative experts within particular activities.

Previous chapters have examined the processes of development. They have included describing activities, and the associated systems of learning that provide the vehicles for expertise. I have discussed how particular sorts of expertise or distinctive sorts of developmental sequence take place within particular 'contexts of use'. The question with which this chapter is concerned is more general.

Stepping back a pace I now ask about the profiles of development within families, given the particular activities that go to make up practices. Families in New Zealand have occupied the foreground in this book. The question is, what is known about the expertise that children from these families develop in literacy expertise? In answering this question I trace relationships with the information that exists on profiles from other countries.

The significance of asking this question lies in filling out the larger picture of how expertise develops within a society. This is an exercise in keeping with the theoretical model in Chapter 1. Understanding the personal processes of learning is enhanced by knowing something about the developmental backdrop.

> The formal education system and researchers contribute to the definition of what counts as expertise.

The question is also worth asking because it tells us something about how a society and its institutions look for expertise. One of the institutions in a society is the formal educational system and another is formed by the researchers who describe and explain developmental and educational processes. Each contributes to the definition of what counts as expertise and what institutions then are prepared to look for.

Contexts and Tests

Unfortunately, this is the briefest chapter in the book because it is a chapter of gaps. Our knowledge in this area is very limited. It is limited in that few studies have systematically described developing expertise in the range of activities described in previous chapters. Partly this arises because of the limitations in resources for, and ingenuity in, carrying out research. For example, we know a lot about the activity of reading books to children at home. This is because it is relatively easily captured by research tools, even in its ordinary daily occurrence. But it is harder to capture the embedded writing episodes, even in something relatively fixed like the activity of writing a name. Consequently, very little work has been done on writing in particular activities.

But the research knowledge is limited in a more fundamental way. The studies are restricted to knowing only some things about reading and writing. Something is known about expertise that is constructed in reading for items, reading for concepts, and writing to name, in particular contexts of use. The particular contexts have been, for the most part, those of formal testing. A frequent practice has been to ask children to write their names by themselves. This procedure has also been adopted by some schools in New Zealand when gaining a profile of children on entry to school (Thackery, Syme, & Hendry 1992).

It has been customary to refer to the resulting knowledge as decontextualised, in that the familiar supports for the knowledge that are provided at home are not present. But it is not reading and writing without a context, even though children are asked to carry out reading and writing tasks unaided and on standardised materials. Formal testing also employs activity settings of sorts, too, in which the participants have intentions which may or may not be shared (for example, to display knowledge for evaluation). They have participation structures and there are limited interactions using particular sorts of tools.

> Formal testing describes the expertise of some children more adequately than others, because children have different degrees of familiarity with this activity.

Questions can be raised about what information gathered in this way from children before school might tell us. The testing will describe the expertise of some children more adequately than others, because children have different degrees of familiarity with this activity. It will capture the strengths of children that are more familiar with the format of independently showing what they can do for evaluation. Some children may be more familiar with displaying their knowledge for someone, even at home (e.g. Heath 1983). Also, testing will represent what children can do well in this activity the more that children understand and share the intentions of the tester. This partly depends on the language that the participants use. Tester and testee will communicate effectively and understand each other to the degree that they are familiar with each other's ways of communicating. In many instances this is dependent on tester and child sharing similar communicative backgrounds because of common cultural group membership.[1]

> Standardised-testing contexts provide limited information because typically they are focused on a narrow band of knowledge.

There is another way in which standardised-testing contexts provide limited information. They are limited because typically they are focused on a narrow band of knowledge. This is another reason why we know very little about the development of the range of activities such as performance expertise and its associated psychological skills. We know little about the development of reading for messages. We know little about the development of writing narratives and expositions before school.

In some of these cases the lack of information reflects beliefs about what sorts of activity are associated with learning effectively at school. This is a rather self-serving or circular sort of argument. That is, the reason some activities, such as reading for narratives, are associated with learning at school is because the school forms of literacy recognise and build upon this expertise. Conversely, children experienced in these activities are able to make more sense of the instructional activities.[2]

Expertise on Entry to School: Concepts, Letters, and Writing

General profile

> In New Zealand, the formal descriptions of expertise have been based on three tests: item knowledge, concepts about print, and how many words children can write.

Three areas for our descriptions of expertise are relatively rich. In New Zealand there are available several 'snapshots' of children's expertise, almost over two generations. The snapshots are based on three tests from Clay's (1979) diagnostic survey. These comprise a test of item knowledge, which involves identification of letters of the alphabet from a list; a test of concepts about print, which probes knowledge of some conventions of print such as orientation of books, and the directionality of print on a specially designed book; and a test of how many words a child can write. Each test requires the unaided demonstration of knowledge.[3]

In 1966 Clay reported results of using the first two measures of letter identification and concepts about print in her seminal study of 100 metropolitan Auckland children. Almost two decades later she reported data from another 72 children who were new entrants at four central Auckland schools and two suburban Auckland schools (Clay 1985). By this time writing vocabulary had become a regular part of the diagnostic procedures too. A colleague, Shirley Nalder, in an unpublished study conducted in 1985, described 16 children from middle-class families in suburban schools. Finally from the late 1980s there are descriptions expertise from children in our SOL study. The data on the three measures of knowledge are shown in Table 8.1 (p. 149). Each of these studies went on to take the same information when children had been six months at school. I return to this six-month comparison later in the chapter.

Table 8.1 Children's average scores on concepts about print, letters identified and words written, in four studies over more than 20 years

Age (yrs & mths)	C.A.P.		Letters		Writing	
	5.0	5.6	5.0	5.6	5.0	5.6
Clay 1966 (n=100)	5.1	10.7	3.9	16.2	-	-
Clay 1985 (n=72)	7.2	13.6	15.5	36.5	2.1	13.9
Nalder 1985 (n=16)	6.8	16.0	13.6	41.6	2.1	21.0
SOL 1990 (n=17)	8.1	14.2	15.2	43.5	1.3	8.6

In general, the more recent studies show that by 5 years of age, children have developed around seven concepts, are able to identify 14 or 15 letters, and can write about two words; typically their name and one other. The concepts about print which have developed show that children generally know how to hold a book, they know where the messages are in a book (in the print rather than the illustration), and they know something about directionality and one to one correspondence.[4]

The studies show that the letters children knew were typically those associated with family names, in both lower and upper case. This knowledge and its basis in family identity is illustrated in the writing of children in the SOL study. Of the 14 children who wrote any word when formally requested between the ages of 4 years 6 months and 5 years, 13 wrote their name. For all 13 it was the first word they wrote. For the fourteenth child it was her sister's name.

A difference of 20 years

What is interesting in the descriptions in Table 8.1 are the comparisons between children's knowledge collected in the 1960s and those collected in the 1980s. It would appear that children entering school in the 1980s knew considerably more letters and knew two or three more concepts about print than children did almost a generation earlier.[5]

What does this difference represent? Among other things, it illustrates the historically bound, but dynamic nature of development. Children's development reflects cultural and family practices which are peculiar to times and places in the development of a culture and a society.

But having said this, it is difficult to tie such a difference to specific changes in our society and family practices. Certainly the studies providing the descriptions cannot shed much light on this question. However, there are some obvious candidates, reflecting both local and international trends. The increased availability of suggestions and guidance for parents for child-rearing tasks is one such trend. Another is the increased availability of early-childhood education.

At the level of family values and beliefs there are international indicators that

> Children's development reflects cultural and family practices which are peculiar to times and places in the development of a culture and a society. It is historically bound, but dynamic.

views have changed about the fostering of literacy knowledge before school. Twenty years ago parents may have hesitated over fostering skills seen to be the responsibility of professional educators (e.g. Durkin 1966). But there has been a shift to parents using the rapidly expanding resources of parent education publications in many aspects of child rearing, including literacy activities (Clarke-Stewart 1978).

In New Zealand, government-sponsored programmes such as 'On the Way to Reading' (see Nicholson 1979) have contributed to the changing views and family practices. The latter programme deliberately sets out to influence the knowledge and beliefs about reading, especially reading practices at school, that parents of preschoolers had in the late 1970s. The vehicle was a radio programme designed by educational authorities which stressed the contemporary meaning- and language-based approaches in classrooms.

In some respects the messages in the programme were those the parents already knew. For example, 99 per cent of almost 1000 parents surveyed said they already spent time in daily reading to their child. In some areas, however, some parents who participated in the programme did report that having heard the programme and participated in group discussions, they had changed their beliefs. Changes included the degree to which parents reported they were concerned about the need for children to learn the alphabet or to learn phonics as a specific prerequisite to learning to read (Nicholson 1979).

Early-childhood education has expanded rapidly over this time period reflected in the profiles in Table 8.1. From 1982 to 1991 the percentage of 2- to 5-year-olds in some form of early-childhood service has nearly doubled (Ministry of Education 1993). These settings provide access to some activities, including reading for narratives and writing names (see the following discussion in Chapter 9).

Each of these sources of influence may have contributed to the changes over time.[6] The significance of the changes, however, are mostly in terms of implications for teaching practice. An implication of this finding is that schools and preschools need to have flexible ways of gaining profiles from their new entrants and 'graduates', which need constant revision. Schools cannot assume that either the range or general level of knowledge of children entering school is relatively fixed. And, given the more general implications for early-childhood education which are developed in Chapter 10, preschools cannot assume fixed sorts of learning and developmental profiles during the time children spend in early-childhood education.

Differences between children

An important finding within the three specific measures described in Table 8.1 is the presence of considerable variability between children. For example, in the SOL study the number of concepts held by different 5-year-olds varied between five and 10 concepts. The variation for letters was even larger, between identifying no letters and identifying all 54 of the test letters. Measured in this way, some children arrived at school with considerable knowledge. Others, even in a group of children predicted to become high-progress children at school, arrived with little or no formal knowledge as measured on this test of independent identification.

In keeping with other educational descriptions, the profiles of different socio-cultural groups also have been monitored. The breakdown for the groups in Clay's (1985) study is summarised in Table 8.2. Clay described children who were from Pakeha, Maori, and a number of different Pacific Islands families.

When expressed as an average, the three groups of children (24 in each group) arrived at school with different amounts of knowledge as measured on the three tests. The most striking of these differences was in the number of letters able to be identified from the list of 54. Less striking, but equally significant in terms of school-based literacy (see Chapter 7) was the number of words the Pakeha children could write unaided compared with the other two groups.

What is not shown in this table, however, is the degree of variation around each of these averages, which was substantial. For example, in the SOL study this variation meant that the differences between children in each group were larger than the differences between the groups. At the extremes was a Samoan child at 5 years who knew 12 concepts about print, identified 23 letters, and wrote four words, and a Maori child who knew six concepts but identified no letters and wrote no words.

Table 8.2 Three groups of children: averages for Concepts about Print, Letters Identified, and Words Written over the first six months of school. (from Clay 1985)

Age (yrs & mths)	C.A.P.		Letters		Writing	
	5.0	5.6	5.0	5.6	5.0	5.6
Maori	6.2	12.6	7.4	34.6	1.2	11.3
Pacific Islands	6.1	12.2	10.6	31.8	1.4	12.9
Pakeha	9.5	16.1	28.5	43.3	3.8	17.5

Individual children had different sorts of highs and lows in their profiles, too. A Maori child identified all letters of the alphabet, wrote one word, and knew nine concepts. A Pakeha child, who could write the highest number of words (five), also knew a lot of concepts (10 concepts) but had a relatively low letter identification (11). The greatest variation was in number of letters identified and number of words able to be written. As all of the families read storybooks almost daily, and they used the narrative activity of reading for narrative (as well as others), the lesser variation for concepts about print may reflect the degree to which this was a common emphasis in each of the families.

These profiles underline the socialisation model of development discussed in this book. In a country such as New Zealand many common sources of experiences of literacy are available to children in ambient and joint activities. Within tutorials and via their personal constructions, children develop knowledge during their years before they start school. Against this background there are notable and important differences between families and between groups in the emphasis they place on certain sorts of experiences.

Profiles such as these, however, under-represent some children's expertise in two ways. The first is an under-representation because of the nature of the activity of formal testing which I noted above. The second is that only a limited

> Children develop knowledge in their preschool years, but important differences occur between families and between groups in their emphasis on certain sorts of expertise.

> That children within the same cultural group can vary as much from one another as they can from children from other cultural groups, directs attention to the sorts of beliefs, knowledge, and messages that families prioritise, and the resources that families deploy for literacy practices.

sort of knowledge base is sampled using this testing context. That children within the same cultural group can vary as much one from another as they can be different *on average* from children in other cultural groups, directs our attention to the sorts of beliefs, knowledge, and messages that families prioritise and the resources that families deploy for literacy activities.

What is most important about this variability is the awareness that labels for cultural and social identity are just that, labels. It is what happens in families that constructs development, not how families are identified for convenient research purposes. The practices and the specific activities that families select, arrange, and deploy for children are what matter for development.

The profiles available from other countries also show significant differences between families, which reflect their respective practices and activities. In London Barbara Tizard and her colleagues tested 343 children from inner-London schools at the end of their nursery schooling, just before entering infant school at 5 years (Tizard *et al.* 1988). The children were mixed ethnically and from relatively low-income families. Their profiles show that, on average, this group of children could name five letters, less than half knew that one reads print rather than pictures, and about a quarter could write their first name correctly.

Again, these researchers found considerable variation. Some children identified all the letters but half could not identify any. A quarter of the children could write their name but almost the same proportion could not write any part of a name. What is different about their profiles, however, is that there were no ethnic-group differences on these measures. But this finding emphasises the point that what happens in families is the important determinant, not a shorthand research label.

The factors most strongly associated with these measures were parental reports of frequent storybook reading using the narrative (or discussion-based) activity; the degree to which parents reported trying to teach reading (incidentally or directly in activities focused on identifying letters of the alphabet) and writing (for example, writing a name); and mother's educational level. These factors were distributed evenly across the ethnic groups.

Descriptions are available also for children from a variety of backgrounds in the United States at an equivalent age but not, therefore, at the point of entry into formal schooling at 6 years of age. On average these children have been found able to name 14 letters of the alphabet and they know several concepts about print. But again, as with Clay's and Tizard's research, the range between individual children in this sort of knowledge has been considerable. The previous chapters have argued that these differences reflect how the practices and activities of families may differ (National Institute of Education 1985).

Development before School

In some respects I have started the descriptions in this chapter the wrong way around. I decided to discuss profiles on entry to school first because they illustrate once again principles about how development takes place, and therefore what differences between children represent. But there is another reason. Even less information is available about development before school.

What is known about what happens before children get to school? Several trends were revealed when development in children's knowledge before school was traced in the SOL study. These trends are shown in Figure 8.1.[7]

The first point to note is that on the measures employed in this SOL study there was a low rate of learning until six months before school, but in the following six months up until school there was a rapid increase in knowledge, particularly letter identification and concepts about print. One source for this increase is parental initiatives. I reported in Chapter 2 how these parents had clear goals for their children's development.

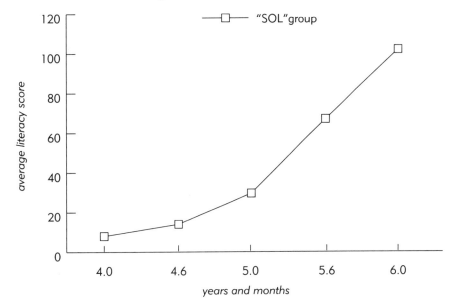

Figure 8.1 The development of children's literacy in the SOL study: a combined count of concepts about print, letters identified, and words written unaided. Note the increase in average score from 4 years 6 months to 5 years just prior to school entry, and the rapid increase after the transition to school

All of the families, Maori, Pakeha, and Samoan, had articulated the general goal to facilitate the children's transition to school. They wanted the children to be able to know the letters of the alphabet before going to school, and they all believed that writing one's name was an important preparatory skill and that all the children should be able to write part or all of their name on entry to school. This suggests an increasing emphasis in families on activities within which this knowledge is constructed. Expertise in writing at 5 years of age was limited primarily to writing one's name.

This burst of activity and its associated spurt in acquisition just before going to school seems widespread. It has been noted in other research in New Zealand (Goodridge and McNaughton 1993) and in the United States (Heath 1983). Megan Goodridge and I found that this activity was focused upon in the last month before going to school. The growth could be sourced to parental beliefs about their children's literacy needs and the goals they set within the activities. Specific activities were intensified, particularly activities focused on conventions and mechanics. Four families in the study bought exercise books with ruled lines in the last month to structure their children's writing with pencils.

> This burst of activity and its associated spurt in acquisition just before going to school seems to be widespread, and could be sourced to parents' beliefs about literacy needs.

But while we found this anticipatory activity to be present in families from different cultural groups the needs for specific preparation were voiced most strongly by Pakeha families, who said that their focus had been influenced by the preliminary orientation visit to the local primary school prior to their child starting school. Heath's (1983) description of this spurt in activities triggered by the impending transition to school was based on white middle-class families.

A second feature of the descriptions in Figure 8.1 is the major differences between children developing in their knowledge over the six-month period of rapid change — the first six months at school. The differences noted earlier between children on entry to school (pp. 150–152) became apparent over this time. At 5 years, for example, one Maori child could identify all 54 letters on the test we used, and one Samoan child could not identify any of them.[8]

Other profiles: activity descriptions

There are other ways to describe children's expertise in reading and writing before school. The procedure which flows from this book suggests that profiles, even for school purpose, should be based on descriptions of activities.

> Preschool profiles, even for school purposes, should be based on descriptions of activities.

A guide to descriptions based on activities is provided by the study by Megan Goodridge and myself, which focused on children's expertise in writing before school (Goodridge and McNaughton, 1993). The descriptions of products were based on the identification of family writing activities within which specific sorts of expertise were located. Activities were defined by a process of interviewing older family members and their 4-year-olds, observing writing, and personal reporting by family members and children. A critical component of this process was a collaboration between the researcher, the child, and the family. In many ways the collaboration took on the appearance of collaborative participation in which valid and reliable (in the sense of agreed-upon) descriptions of the child's expertise were the focus.

The collaboration was active. Parents at first did not provide much in the way of elaborate descriptions of goals or other features of activities and tutorials. This was because these were relatively automatic and embedded in everyday life. The asking and probing activated more deliberate attention to writing activities — yielding over time extensive comments, descriptions, evaluations, and reflections. A common response from families at the end of our first visit was one which expressed some amazement at the range and extent of their activities and what their child seemed to be able to achieve.

The resulting identification of eight activities was based on descriptions of goals, functions, and interaction patterns derived from the information supplied by the 18 families. The categories are shown in Table 8.3 (p. 155), which is a more extensive list than the set of activity systems described in Chapter 7. The research purpose was to provide a more fine-grained analysis which could be closely tied to family ideas and goals.

Two activities focus on conventions: writing letters (of the alphabet) and writing numerical symbols. The six categories of textual activities extend those introduced in Chapter 7. For example, the category of exposition was separated

Table 8.3 Family writing activities with examples of goals and products and the average number of products in each category provided by 18 families over six months. (from Goodridge & McNaughton 1993)

Activities	Goals and Products (samples)	Average Number of Products
Conventions		
Alphabetic	demonstrate knowledge e.g. line of letters	3.9
Numerical	note significant times e.g. calendar date	1.4
Texts		
Naming	identify self e.g. write name separately	11.9
Narration & description	representation of experiences e.g. personal account	10.3
Exposition	locate self in neighbourhood e.g. writing parts of map	6.0
Labelling	show ownership e.g. write name on product	4.4
Messages	express emotion to others e.g. letter to family	2.4
Diverse	experimenting with categories e.g. combined name, numbers, & labels	3.6

from the more general writing activity of narration, description, and exposition. Additionally, a category of labelling was identified to capture the specific use of names and single words to label products (samples of writing). The category of diverse activities covered those activities which were hard to categorise because they seemed to be in transition and a mixture of other categories.

The identification of these activities provides a different profile of expertise from the ones summarised in Table 8.1 (p. 149) and Table 8.2 (p. 151). Over 800 products representing these eight different categories of activities were collected from the 18 families involved. A summary of the respective numbers of products from these activities is shown in Table 8.3. All but two of the families produced products from five or more of the activity categories.

> These descriptions suggest that children's emergent expertise on entry to school is extensive.

These descriptions of activities suggest that children's emergent expertise on entry to school is extensive. The research process also suggests that obtaining adequate descriptions of that expertise requires considerable familiarity and knowledge gained in collaboration with the family.

Development over the Transition to School

Not surprisingly, once children go to school there is a rapid increase in the expertise that is measured on the tests of letter identification, concepts about print, and writing. Figure 8.2 (p. 156) summarises the data from Clay's (1985)

Figure 8.2 The expertise-development profiles of the three groups of children in Table 8.2 (from Clay 1985). Note rates of change over six months

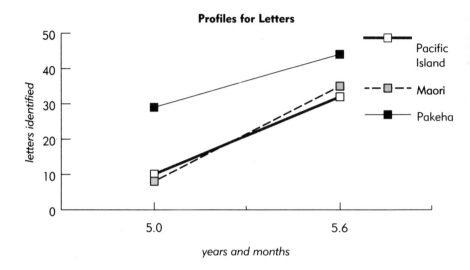

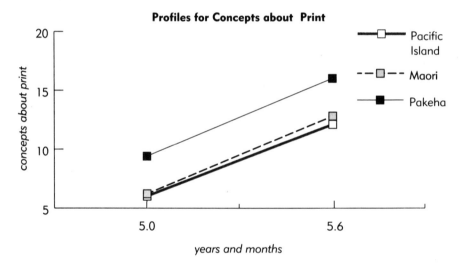

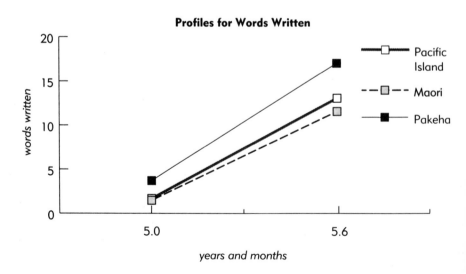

study, depicting this accelerated learning from 5 years to 5 years 6 months. It is not surprising, because schools deliberately set out to teach these forms of expertise. And the first two tests (letter identification and concepts about print) tap a limited set of items or concepts; technically, these have ceilings and children rapidly reach the ceiling and can learn no more. By 5 years 6 months children in the recent studies (see Table 8.1, p. 149) identify 30 or more of the 54 letters, and the differences between groups have reduced in overall terms.

This rapid increase is generally found in studies that have followed children over the transition to school. Another trend found in such studies is the relative advantage of high-scoring children. The advantage comes in the continued higher levels of knowledge as measured in these sorts of tests, and even in accelerated rates of learning. The ceiling effects on the simple measures of letter identification do not show this. But even over six months the potential for a widening gap is suggested by the measures of C.A.P. and writing words. The accelerated learning has come to be known as a 'Matthew Effect', in which those children who have more expertise which is functional come to obtain increasingly more out of their schooling (Stanovich 1992).

> The 'Matthew Effect' describes how children who have more expertise which is functional in classrooms, come to obtain increasingly more out of their schooling.

The scores on these sorts of tests are related to further progress at school. But they may have different relationships at different instructional periods of learning to read and write. For example, letter knowledge is highly related to early progress in reading and writing at school. The expertise in reading for narrative and writing a name has been more closely linked to later achievement in reading and writing at school, particularly when there is increased focus on comprehension of and writing of narratives (see Snow *et al.* 1991; Tizard *et al.* 1988).[9]

Summary

Children develop particular sorts of expertise during their preschool years at home and in other socialisation settings. There are profiles of a restricted range of forms that this expertise takes. These are the sorts of expertise that are effective as preparation for school instruction. Families vary in their practices which are related to these forms of expertise.

> Children's expertise is bounded by time and place and culture. It is situated within particular activities and reflects the focus and range of those activities.

I have suggested by descriptions of its early development that children's expertise is bounded in several ways. It is bound by time and place and culture. More specifically, expertise is situated within particular activities and reflects the focus and range of those activities. This is summarised in the following proposition.

Proposition Six: What children learn to do with written language is become relative experts within particular activities.

The question which I pursue in the last two chapters is why these relationships exist between expertise developing before school and at school, and what the educational implications are of finding relationships. These are not inevitable sorts of relationships. Families do not need to be at a relative 'disadvantage'. And the relationships with school do not reflect a single developmental path along which all children proceed but at slower or faster rates. On the one hand

Multiple literacy practices are possible, as are previously unrecognised forms.

the variations between children reflect family literacy practices which are not fixed. Multiple literacy practices are possible. On the other hand, the ways in which schools might build on forms of literacy can be modified so that forms previously unrecognised by teachers are effective, too.

Implications

For families, educators, and researchers

Families: The major implications for families are discussed in chapters 9 and 10. One point can be made here. Clearly, children in all families in a heterogeneous industrialised society such as New Zealand are developing expertise (including, of course, knowledge) in family literacy activities before school. The implications concern the possible roles in fostering different forms of expertise that families may choose, or be able, to assume.

Educators: Again, the major implications of this chapter for educational practices are the focus of the next two chapters. The implications all derive from the question of how educational practices interface with systems of learning and development in families.

There are major questions about how educational practices interface with family systems.

The research agenda is to analyse activities across settings, and gather profiles which can be linked with family identity.

Researchers: A substantial research agenda is introduced in this chapter. It is to analyse activities and gather profiles which can be linked with family identity. There are theoretical challenges to develop ways of describing and explaining activities and variations in these within and between families, and within and across home, early childhood, and school settings.

Further Reading

General profiles:

Tizard, B. Blatchford, P., Burke, J., Farquhar, C. & Plewis, I. (1988). *Young Children at School in the Inner City*. Lawrence Erlbaum Associates, London.

Snow, C. E., Barnes, W. S., Chandler, J., Goodman, I. F. & Hemphill, L. (1991). *Unfulfilled Expectations: Home and School Influences on Literacy*. Harvard University Press, Cambridge, MS.

Methods for describing knowledge and activities:

Clay, M. M. (1979). *The Early Detection of Reading Difficulties: A Diagnostic Survey with Recovery Procedures*. Heinemann, Auckland.

Dickinson, D. K., De Temple, J. M., Hirschler, J. A. & Smith, M. W. (1992). 'Book reading with preschoolers: Coconstruction of text at home and at school'. *Early Childhood Research Quarterly*, 7, 323–346.

Tharp, R. G. and Gallimore, R. (1988). *Rousing Minds to Life: Teaching, Learning and Schooling in Social Context*. Cambridge University Press, Cambridge.

End of chapter notes

1. These points about the limits and discriminations of testing are discussed in a number of sources (e.g. Donaldson 1978; Erickson 1984; Newman Griffin & Cole 1989).

Sociocultural perspectives such as the co-constructivist framework adopted in this book offer alternative possibilities for assessing children's expertise. Gaining a profile of children's forms of expertise might well include describing what children can do in the formal testing setting (Brown & Palincsar 1990).

This provides information on the one hand about the generalisation of expertise to this setting. On the other hand it also provides some information about expertise in relationship to the contexts of use favoured by school. But Bronfenbrenner's (1979) concern for 'ecological validity' should be kept firmly in mind. That is, testers need to know the relationships between what they are describing in such a test, and the expertise (and activity settings) about which they then make judgements.

Clearly, the history of testing is one of incipient discrimination against children and their families for whom the activity is less familiar. The history of this debate in New Zealand can be seen in a series of articles on the uses on a scholastic achievement test (TOSCA) published in the New *Zealand Journal of Educational Studies* 1987–1988.

2. It is possible, however, to develop informed instructional bridges between home and school activities (these are discussed in chapters 9 and 10). One of the consequences of effective connections between these settings is the likelihood that under the circumstances of more effective bridges for diverse groups of children the strong correlations between a narrow range of literacy activities and school achievement in reading and writing would weaken.

3. Marie Clay developed a number of tests which provide a diagnostic profile of children after a year at school (1979). The profile is based on markers of progress in the school system of New Zealand although it has been adapted for use in other countries too. It is used to make judgements about rate of progress and about the need of a child for remedial intervention. The tests use an independent demonstration format. In addition to alphabet knowledge, concepts about print, and the writing test there is a procedure for taking a record of strategies used when reading texts.

4. This general profile is supported by descriptions in a number of other studies. It appears that the majority of children come to school knowing some letters, able to write their name, and knowing several concepts about print (see Hendricks, Meade & Wylie 1993); Phillips 1986).

5. The differences over the 20-year period are not obviously related to differences between the types of families in the samples. Although the more recent small-scale studies, such as Nalder's and SOL, used selected groups of children, the larger-scale studies by Clay (1966; 1985) used representative samples from urban and suburban areas.

6. It is interesting to note that the expansion of television over the 20-year period (1960s–1980s) was not associated with declining knowledge of how books work and other forms of literacy knowledge. Under some conditions moderate television viewing, for older children, can have some beneficial effects on aspects of reading (Anderson, Wilson and Fielding 1988), especially under the circumstance of written language accompanying television images (Elley 1992). And evaluations of children's television have shown beneficial affects on item knowledge in particular (see National Institute of Education 1985).

7. Exact data points (at the age intervals) for each child at each age have been approximated from the lines drawn on graphs for individual children. In the three cases where parts of the data were not available for children before 4 years 6 months, the first data point was interpolated (drawn back using the trend in the later data points as a guide).

Because of some individual gaps in the measures the numbers of children at each age level vary; for this graph the average numbers are based on 17, 17, 16, 15, and 15 respectively.

8. The descriptions available in other studies also show emerging differences between children during this time (see Hendricks, Meade & Wylie 1993; Phillips 1986).

9. The nature of developmental processes needs to be reiterated here. The relationships between narrative expertise and writing a name and later achievement at school occur because these interact with other activities and processes continuing at home and at school. An inoculation model for development is not appropriate. Teaching a child to write their name at the age of 4 will not necessarily affect achievement at school.

Chapter Nine

Settings: home, early childhood, and school

> **Focus**
>
> **Development and the relationships between settings**
> - Different settings within which children develop can be well or poorly co-ordinated.
> - Co-ordination is dependent on whether practices, activities, and systems of learning and development are similar or complementary. Whether these are similar or complementary also depends on a number of other concerns, including how the participants in each setting relate to one another.
> - Development is enhanced by the degree to which children's settings are co-ordinated.

The activities in which children participate provide the structure for their development, and development increases participation in socio-cultural activities.

One idea underlies much of what has been said in the previous chapters. It is that it makes very little sense to separate the study of what develops from how it develops. I have argued that the activities in which children participate provide the structure for their development. So to understand development one has to understand activities, and vice versa. Barbara Rogoff (1993) puts this idea very clearly when she says that development is increasing participation in socio-cultural activities.

The basic argument, that developmental processes and products reflect and build upon each other, has been repeated in this text in several ways. It occurred in the analysis of expertise within tutorial systems (see Chapter 4). A similar argument was made about learning systems within activities; the activities create a framework for learning and the processes of learning contribute to and alter that framework over time (see chapters 5, 6, & 7). At the most general level of analysis, activities and learning systems are not divorceable from families in which they take form. They come into existence because of the resources that families deploy, and they contribute to those resources. They express and construct values, beliefs, expectations, and knowledge for family members.

There is a further level of analysis. Families do not exist in a vacuum. Families live in social, cultural, political, and economic worlds, strengthened and buffeted by the elements of those worlds. For example, the availability in the child's home of people with whom a tutorial system can develop is dependent on the dynamics within the family. And this availability is in turn dependent, among many other things, on patterns of work and how much time family members might have to do things, such as read books with their preschoolers.

This level of analysis must include more than the family. The development of children's expertise does not take place only in family settings. Other settings contain opportunities for ambient, joint, and play activities. When these are deliberately organised, as they are in formal educational settings, complex developmental processes come into play. Not the least of these are the relationships between the forms that literacy takes at home and those in the other more deliberate settings. These relationships raise important questions to which I have referred in previous chapters. Does it matter to their achievement at school or learning in early childhood settings what a child is (or is not) learning at home? Conversely, does what they are learning in an early-childhood setting or school matter to development at home?[1]

> Other settings, beyond the family, also contain opportunities for ambient, joint, and play activities.

This chapter and the final chapter explore the development of literacy using the family as the focus. I look at the way the family is situated in contexts such as the neighbourhood and the worlds of school and work, and how these combinations of contexts contribute to literacy expertise. Although this book is primarily about literacy in the family, this chapter examines the transitions to preschools and school, and the relationships between the family and these educational settings. Ways of understanding relationships between learning and development at home and in formal educational settings are explored. This chapter deals with concepts and explanations, enabling me, in Chapter 10, to draw some implications for educational practices for children at early-childhood and beginning-school levels.

Families in Context: Bronfenbrenner's Model

> To understand how the family lives and develops in various contexts, a way is needed of describing them as part of the larger economic, political, and social worlds.

In order to understand how the family lives and develops in these various contexts a way is needed of describing it as part of the larger economic, political and social worlds. That way was introduced in Chapter 2, through the concept of the family as a system. Although this concept has been used to describe the properties of a child's learning (with others and independently), it can be applied to this wider, more complex grouping. A family is dynamic; it grows and adjusts to various forces, and is goal oriented.

Describing the family as a system enables us to see it as situated in other encompassing systems. The notion of systems within systems comes from a model of human development provided by developmentalist Urie Bronfenbrenner in 1979. It is a model which conceives the contexts of development in terms of nested systems. The model is represented in Figure 9.1.

> Microsystems involve the child in interactions with significant others in the immediate environment.

The family contains what Bronfenbrenner called **microsystems**. These are the basic vehicles of development involving the developing person in interactions with significant others in the immediate environment. Microsystems allow what Bronfenbrenner called primary developmental contexts and secondary developmental contexts to occur. The former are developmentally progressive interactions with those with whom one has a positive emotional bond. The latter are opportunities to play and explore which develop from the former.

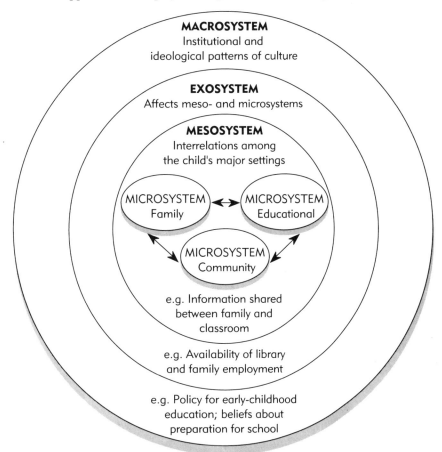

Figure 9.1
Bronfenbrenner's ecological model: the family within enveloping ecological systems

These original ideas of intimate and progressive developmental systems have been incorporated into the concepts used in this book. They are employed in the more focused concept of systems for learning and development (tutorial and personal systems) which take form within specific activities. In this book the psychological explanation of how development takes place in these systems has stressed processes of co-construction.

Families are a context for activities and their associated systems which create learning and development. Families do not have a monopoly on learning systems. They exist also in other settings within which the child has an important

> The psychological explanation of how development takes place in microsystems stresses processes of co-construction.

role. Before school this includes early-childhood education settings. It also includes significant institutions such as Sunday school, church, and children's clubs. In New Zealand it includes other cultural settings, such as the marae (the meeting place for the family, sub-tribe, or tribe to gather, and which comprises buildings for formal meeting as well as for eating and sleeping and eating).

Grouped together these microsystems constitute a second level of context. Microsystems are nested within a wider system which Bronfenbrenner called a **mesosystem**. The identification of a mesosystem provides a means of describing and developing explanations about how development (and learning and development systems) are affected by relationships, interactions between settings. The relationships form at the point of making a transition from one setting to another, and from that point in ongoing relationships between settings. Bronfenbrenner regarded transition points and ongoing relationships between settings as particularly significant for development and proposed that:

> The developmental potential of a child rearing setting is increased as a function of the number of supportive links between that setting and other contexts involving the child or persons responsible for his or her care. Such interconnections may take the form of shared activities, two way communication, and information provided in each setting about the others. (Bronfenbrenner 1979, p. 847)

> Grouped together, microsystems are nested within a mesosystem, which provides a means of describing how development is affected by relationships between settings.

This proposition about mesosystems is central to the following discussion. Bronfenbrenner's model of developmental contexts did not stop there, however. He went on to describe how mesosystems in turn are nested within an **exosystem**. The exosystem describes the settings and agencies of which the developing child is not a member, but which impinge on the mesosystem and the microsystems. It includes the worlds of work, the neighbourhood, resources and agencies including health, welfare, and education, and the political system. This level of analysis was employed in Chapter 2.

The final level at which developmental contexts can be identified was given the term **macrosystem**. Bronfenbrenner described this as the overarching beliefs and practices of the cultures and subcultures with which the family identifies. I have described aspects of these elements and how they influence family practices of literacy in previous chapters. For example, it is clear that different pedagogies carry cultural messages. Reading for narratives is a different cultural act from reading for an accurate performance (Chapter 6).

Transitions and Connections

Figure 9.2 (p. 165) is a representation of the transitions that a child may make from their family to other settings over their first few years. The transitions that are of concern here are those that carry the potential for Tutorial systems and Personal systems to develop. Three such transitions to different settings are shown: the transition to early-childhood educational settings (Transition 1), to other significant institutions, such as church and marae (Transition 2), and to school (Transition 3).

These particular transitions are influential. So too are the ongoing connections that then develop between these settings over time. They constitute the fuel of Bronfenbrenner's mesosystem. The diagram shows these connections, focusing on those between the home and each of the others: Connection 1 — between the family and early-childhood settings; Connection 2 — between the family and other institutions; Connection 4 — between the family and school.

It is, of course, possible to examine the connections that exist between settings outside of the home. The connection between an early-childhood setting and a church is an example. But given the focus of the book I do not explore all these possibilities and, except for one, they are not illustrated in the diagram; that is the connection between preschool settings and the school (Connection 3). I discuss here the properties of transitions and connections that are likely to facilitate the development of the multiple forms of literacy that an educational system should value.

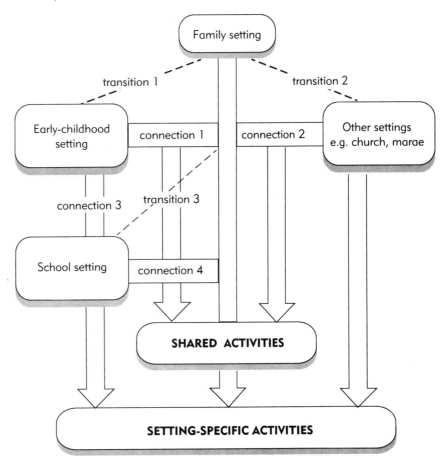

Figure 9.2 Transitions and connections between the home and other developmental settings

Figure 9.2 also reflects an important characteristic of development that has been a major point in earlier chapters. The development of expertise takes place in each of the major settings, and the relationships between activities in each setting may vary, that is, the activities in which children engage can be relatively distinct to a setting, or can be shared or complementary.

The Case for Matched Settings (and Some Implications)

Considerable research over the past 20 years bears on the prediction that Bronfenbrenner made about close connections. In one strand of research a 'continuity' or 'match' hypothesis has been proposed. It is used to explain why children from some cultural groups, particularly those described by Ogbu (1991) as 'involuntary minorities', and first peoples who have experienced colonisation, are not well served by 'mainstream' schools. The argument has been that beliefs about schooling, patterns of language use, and forms of learning that have developed outside of school do not match those at school. The psychological processes entailed in this match are both personal (what sense the learner can or chooses to make of the new setting) and interpersonal (how easily shared goals develop within school activities).

A number of educational interventions in classrooms have attempted to match culturally based patterns of discourse and pedagogical styles used by children outside of the school. Systematic analyses of these interventions generally show that when discourse patterns and pedagogy are well matched children are more likely to achieve at higher than typical levels (for example, in reading), at least in the short term. But some boundary conditions apply. These are most likely to occur where the children's community identifies with the school, perhaps through assuming some control or functional participation in school decisions and activities (see Cazden 1988; Cummins 1986; Tharp & Gallimore 1988).

The most extreme form of matching can be found in those schools that have been created and controlled by a community, and that function as linguistically, pedagogically, and (therefore) culturally distinct. Such radical alternatives to mainstream schools have been designed in various countries to meet cultural and political goals within education. In New Zealand Te Kohanga Reo (literally, language nests) and Kura Kaupapa Maori (Maori-immersion elementary schools) represent this development. They can be extremely effective sites for cultural construction and educational achievement (Smith, Smith, & McNaughton 1989; Hohepa *et al.* 1992; McNaughton 1994a).

A second form of intervention research has worked primarily to influence the relationship between activities at home and school from within the family setting. These interventions set out to increase the presence of activities occurring at home which are related to classroom-based activities. For example, parents have begun to hear children reading appropriate instructional texts using tutorial patterns that are consistent with those used at school (Hewison 1988; McNaughton 1987). When focused on the home–to–school transition, such interventions have increased reading to children in ways that construct the activity of reading for narrative described in Chapter 5 (pp. 89–92) and Chapter 6 (pp. 118–119) (Sulzby & Teale 1991).

Again, what has emerged from this research is the recognition that important boundary conditions apply for anything happening at home that affects classroom learning. These conditions include the degree to which the parents and teachers know what each other is doing. It also includes the degree to which the activities are, and continue to be, similar and progressive (McNaughton *et al.* 1992b).

> Research about close connections explains why children from some cultural groups are not well served by 'mainstream' schools.

> With well-matched discourse and pedagogy children are more likely to achieve at higher than typical levels.

> The most extreme form of matching occurs in schools created and controlled by a community, and that function as linguistically, pedagogically, and culturally distinct.

These two lines of research and their associated arguments of close communication, power, and control lead to a very significant educational challenge. How can settings which are harmonious and complementary be created for different communities? How can settings in which multiple developmental pathways are promoted and where multiple forms of literacy are recognised be created? Can settings enable children to become expert at 'reading' and 'writing' within the activities which they confront at home as well as those needed at school? In Chapter 6 this was identified as an ideal (third) form of intervention, that is the creation of joint settings which promote ways of contributing to forms of literacy both at school and out of school.

Connecting Settings

Bronfenbrenner made a strong prediction about the developmental significance of connections between settings. This significance for a child's learning is explained by psychological processes involved in co-construction. Children's learning can be explained by the interactive and constructive processes that take place within and across activities through learning and development systems (e.g. at home and at school).

> If activities are similar across settings then expertise is able to be practised more extensively.

In the first instance, if activities are similar across settings then *expertise is able to be practised more extensively.* The power of practice is harnessed if complementary personal systems develop across settings. The presence of problem solving and analysis focused on similar activities creates a bridge between the worlds of home and other socialisation settings. Further learning occurs if tutorial systems which build upon expertise develop also. The learning that becomes possible under these circumstances is multiplicative, that is, the addition of an extra setting results in more than a fixed amount of further experience.

> Accelerated learning occurs because the development of expertise acts to increase the basis for further learning within activities.

Because children are active constructors and because processes of self-regulated learning occur within personal and tutorial systems, opportunities for learning are accelerated when settings and their activities are connected. Accelerated learning also occurs because the development of expertise acts to increase the basis for further learning within activities. If a child learns a new concept about books, such as where the message is encoded (in the patterns of symbols, or letters), the potential for making further discoveries (for example, about the invariant properties of these symbols) is vastly increased.

How are connections which foster these processes made? Close connections and transitions are made at two levels. One is through the action of the learner as he or she moves between settings, playing, practising, and learning with expertise. This process of transporting expertise could be described as **generalisation** or **transfer across settings**. But this does not mean that it happens in a vacuum. The second level is the co-constructing of properties of the settings as developmental contexts. A similar activity system at school, such as reading storybooks to children (e.g. in shared reading), creates a bridge with the activity system at home.

These two levels are interdependent and each determines the other. At one level it is the learner acting within activities across settings that determines whether what happens in one setting bears on learning in another. But the learner's actions are co-constructed (e.g. with family members and/or teacher) in activities which define the purposes for literacy practices in that setting. So the functions of activities across settings also play a part.

The ideas that each participant (family members, teachers, and child) has about the activities (including the goals and tutorial processes) are important parts of the connecting process. These ideas channel (direct) their actions. For example, parents hold beliefs about why a child might ask how to write their name. These include whether this is an appropriate thing to learn and how one shows or teaches it. These ideas constrain what a parent then does. For each participant, ideas have this channelling role. The channelling of actions takes place with parents, siblings, the child learner, and significant people in the other socialisation settings that the child enters.

So, for example, it matters whether or not an educator in a preschool setting conceives of writing one's name as an important educational task. It also matters whether or not they believe that part of the educator's role is to communicate such developmental expectations to the child's parents.

In this way, making connections between settings is not unlike the process that underlies co-construction within activities at home. In the case of meso-systems, it is between groups of people and activities across different settings. It is the establishment of familiarity through a joint focus and the establishment of intersubjectivity (shared goals and ideas about activities).

Contributing to the co-construction is the recognition by significant people across settings of new or different ways a child is developing of being expert. For example, a parent suddenly notices that their child can write the first letter of their name because the attempt of this writing is on the painting brought home from the preschool, and conversations with the child and the educator reveal that this is indeed the child's work.

This enables a child to assume their learning has wide significance, that is, if an activity at school is seen to be valued at home and vice versa, then the seeds of a generative (productive) familiarity are sown. One assumes that one's ways of learning and using written language are valued and have meaning for a range of settings within which one lives. Even more significantly, the presence of a disjunction or separation between the systems (e.g. of home and of school) is avoided. This familiarity in turn creates opportunities for interactions in which the goals are shared and interactions are closely adjusted to meet those goals. This means exchanges are attuned to the personal needs of the learner, creating the properties of effective tutorial systems.[2]

> The development of a joint focus in activities and the benefits of a learner assuming that their learning has wide significance are dependent on the actions of people in the two or more settings.

The development then of a joint focus in activities and the benefits of a learner assuming that their learning has wide significance are dependent on the actions of people in the two or more settings. Two significant features of what educators do in educational settings and what parents do in home settings are particularly important to note. These two features (ideas and strategies) are among those that are used to describe someone's expertise. In a sense the development of a shared focus between home and educational settings is an outcome of expert actions by educators and caregivers.

Parents and teachers as expert educators

Ideas

The actions of parents and educators in child rearing and teaching flow from their ideas about their roles. These ideas, in turn, are modified in the light of their actions. The ideas that I am concerned with here are those which contribute to the relationships between settings. They come in three forms. I start with the ideas held by educators.

Ideas held by teachers

One set of ideas involves the beliefs teachers have about their professional role. The most significant concern their roles in helping to create complementary activities between settings. In turn, this is related to their theory of the nature of learning and development. Teachers' actions are influenced by whether they believe relationships between settings influence children's development and whether their role includes contributing to the creating of such relationships.

> Teachers' ideas concerning the educative role in creating complementary activities between settings is influenced by whether teachers believe relationships between settings influence children's development and whether their professional role includes contributing to such relationships.

The second set of ideas is about what the individual child can do and the activities within which he or she can do these things. Teachers have complex procedures for representing this knowledge for school-based expertise. Curriculum statements and pre-service and in-service education are major sources for these concepts. But how much does a teacher know about the child's developing expertise and ways of learning outside of the preschool or school? Like their beliefs about their professional role, this knowledge is theoretically based. Being able to recognise what a child can do depends on what one is prepared to look for. It is a knowledge based on familiarity with activities (including learning systems) and forms of expertise in the other setting (home).

> Being able to recognise what a child can do depends on what one is prepared to look for. It is dependent on a knowledge of and familiarity with activities and forms of expertise in the other settings.

The third set of ideas relates to how one teaches and how children learn, including knowledge of the forms and functions of tutorial configurations. These concepts set boundaries and define channels for developing ways of teaching and learning. Interestingly, the ideas about how teaching and learning take place also apply to how one might communicate to instruct an adult. The ideas about instructing parents will determine how one might convey information to a parent.

Ideas held by parents

Each of the three sets of ideas above are influential at home, too. Caregivers have ideas about their efficacy and role in relating to educational institutions. An interesting example of beliefs affecting decisions about literacy activities comes from Durkin's (1966) study of children who entered school already able to decode some words. One group of families was under-represented in the sample of children who could read on entry to school. These were described as upper middle-class families. The caregivers in these families had decided to actively ignore or even dissuade their preschooler from being interested in reading and writing, on the grounds that if they as parents responded to children's interest, they might do the wrong thing. Their strong belief was that teaching reading was the responsibility of the professional educator.

Ideas develop and change within communities and with repeated experiences. Parents in communities that have experienced a lack of economic and

> Parents in communities that have experienced a lack of economic and political power and a history of schools failing to meet their aspirations develop pragmatic beliefs about relationships with school systems.

political power and a history of schools failing to meet their aspirations develop pragmatic beliefs about relationships with school systems. Their beliefs, which represent common experiences about the inability of school systems to meet their needs, are not conducive to developing close relationships with schools (e.g. Laosa 1989; Ogbu 1991).

Parents also develop ideas about their children's expertise and the activities within which this develops. The joint activities that family members engage in with children are theory driven. They have reasons for organising activities, ways in which members interact, and ways of recognising expertise, sequencing tasks, and knowing when to help; all of these reflect theories about the nature of activities and children's learning. The theories may not be very explicit. Because they are so embedded in everyday life they may be unremarkable, but they are systematic forms of knowledge.

Even simple and non-scientific theories change, develop, and are modifiable. One caregiver's theories about a child may be different from those of another caregiver. This could be a neighbour, but, even more significantly, it could be the professional educator. These two points are critical. The converse of the questions asked about teachers' knowledge (p. 169) can be asked about caregivers' knowledge. How much does a caregiver know about expertise and activities in the educational settings? How much is known about the forms of teaching and learning within these activities?

> Being prepared to see and know about the other setting determines the caregiver's specific knowledge. Familiarity underlines this preparedness.

As in the educator's situation, being prepared to see and know about the other setting (in this case, school) determines the caregiver's specific knowledge. Similarly also, familiarity underlies this preparedness. Some parents know more about school forms of expertise because of their own schooling experiences, because of who they are. For example, having been longer at school, having relatives who are educators, having access to educational groups, all help create greater familiarity. This is one sense of the idea of cultural capital (Chapter 2).

Strategic action

A general feature of being an expert is the development of strategies for acting effectively in particular contexts and for regulating their performance. The form and effectiveness of one's strategies and the degree to which one can monitor them depend on what one knows. This includes the sorts of knowledge I described above involving the goals, expectations, and beliefs about settings and activities, and about teaching and learning applied to individual children.

Teacher strategies

> Enhancing connections between home and educational settings requires educational strategies for communicating with families and checking the effectiveness of one's teaching.

Being a skilful teacher who acts to enhance connections between home forms of literacy and those in the educational setting also entails strategies. These include strategies for communicating with and learning from family members. They also include ways of checking knowledge and the effectiveness of one's teaching based on that knowledge.

Parent strategies

Caregivers also develop strategies. These include constructive imitations of the actions of professional educational experts. An example of this happening is illustrated in the transcript below. It comes from a Samoan mother reading to her 4-year-old daughter at home. Our practice in the SOL study had been to carry out the testing of children in the home, so the testing could be observed by members of the family. Our concern was to have the procedures open and able to be discussed. The Samoan mother had typically read storybooks with her daughter with the standard narrative focus as conveyed in audiotape recordings. After one visit in which she observed us testing for concepts about print we found the presence of different exchanges. She had incorporated our testing questions into the activity ('*Show me where the front of the book is.*').

Sample dialogue 9.1
Modified exchanges following a research visit

MOTHER   First you can show Mummy where the front of the book is? Show me where the front of the book is.
CHILD   Yeah.
MOTHER   The front, where's the front?
CHILD   Here.
MOTHER   Okay. You turn the page. Where's the back of the book?
CHILD   Here.
MOTHER   Where do you start reading? Which side? Which side do you …
CHILD   This side.
MOTHER   Which one?
CHILD   This side.
MOTHER   That's right. Oh! [page turned] This is where the story begins. 'No more cakes.' That's the … that's the name of the story.

Other examples from the same session were described in Chapter 5. This experience made us very aware of our power as researchers in that we held privileged knowledge. But it also underlines the motivation of parents to acquire informed ways of child rearing.

Other parent strategies include those for gaining and conveying information about classroom activities and expertise. Some parents are more likely to approach classroom teachers and ask questions about their child's progress than others (see Chapter 10). In general, such an approach provides opportunities for connections to form and information to become available. Sometimes, however, it may be strategic to do very little in deliberately seeking information. This occurs in circumstances of a close match, when a strong continuity develops between the classroom and the family setting. It may also occur as a strategy of resistance when the parents' concepts of the role of the school suggest that the family has little power and effectiveness in influencing school (Laosa 1989; Ogbu 1991).

> Other parent strategies include those for gaining and conveying information about classroom activities and expertise, e.g. asking about their child's progress.

Relationships between ideas and strategies

The relationship between ideas and strategies depends on a number of factors, such as the available theoretical knowledge and research, and is not automatic. Complex relationships are notable in the case of teaching practice. In New Zealand schools there is widespread knowledge of the general principle of emergent literacy available from Marie Clay's writings (e.g. Clay 1979) and from others. But in the specific area of emergent writing only recently have theoretical concepts become available to schools that emphasise that writing develops before school.

A recent New Zealand Ministry of Education publication, *Dancing with the Pen* (Learning Media 1992), adopts two basic themes: that writing develops before school (*Children ... will have scribbled, painted and made their marks on paper*), and that writing takes many forms. The publication provides descriptors of the characteristics of 'emergent writers' reflecting concepts that the development of writing at school entails affect, purpose, procedural knowledge, and regulation. It was not the purpose of that book to develop an extensive understanding of emergent writing before school and consequently it devotes little time to discussing activities, expertise, and how teachers might 'look' for these.

It is not surprising, then, to find that a recent Ministry of Education survey of the monitoring practices of 100 schools (Thackery, Syme, & Hendry 1992) indicates that little information is collected about writing activities and expertise on entry to school. A minority of schools do systematically collect information, reflecting a view that writing develops before school, but it is a view of a developmental sequence focused on only a few forms of expertise, notably writing one's name or any other words in an independent context not within structures that might be similar to the familiar activities happening at home.[3]

> Whether or not teachers gather information about different forms of expertise in writing depends on their beliefs and knowledge, which are in turn dependent on the availability of research and theory.

Whether or not teachers are interested and act strategically to gather information about situated expertise in writing depends on their beliefs and knowledge, which are in turn dependent on the availability of research and theory accessible through pre-service and in-service education.

The processes involved in connecting settings that I have just explained predict significant roles for both school and early-childhood settings in the development of literacy at home. The roles are both direct and indirect. A recurring example of the indirect role which early-childhood educational settings have highlights the significance of the relationship between these educational settings and homes, and raises issues about the need to be more direct or explicit in this relationship. The example has arisen in a number of studies both in New Zealand and elsewhere (e.g. Heath 1983).

Figure 9.3 (p. 173) summarises the example and explains the roles from the parents' point of view. The process starts with the strategic action of a parent in selecting a preschool for their children's socialisation. Choices reflect the ideas and values they hold for child rearing. Choices also reflect the availability of early-childhood education and the family resources for accessing this form of education.

> Selecting a preschool reflects the ideas and values that parents hold for child rearing, as well as the availability of early-childhood education, and the family resources for accessing this form of education.

Children develop forms of expertise in the early-childhood setting (home). The example which comes to the notice of some parents is writing (initially copying) names under paintings and other creative products. Some parents report how there comes a point at which they see and value this as a newly developing form of expertise. In so doing, their ideas may be modified and their subsequent strategic action at home may change. They may incorporate the naming activity into their practices, whereupon this becomes a shared activity contributing to the connections in the mesosystem.

The opportunity for the educator to have a more direct role exists at another point. That is, parents may gain ideas and strategies because teachers share these directly and explicitly. For example, information can be shared with caregivers about educational activities.

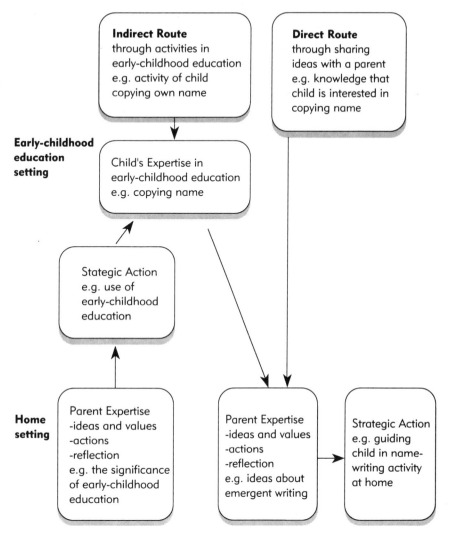

Figure 9.3 The roles of early-childhood education in parents' ideas and socialisation of literacy. Two possible routes through which early-childhood education might interact with the development of literacy activities at home

Peers and the Development of Connected Settings

Connections between the family and educational settings can be built by the peer group, too. Educators are not the only source of activities, and the peer group also creates a Zone of Promoted Actions within which specific sorts of activities occur (see pp. 17–18). This creates the potential for Tutorial systems among peers and Personal systems developing from ambient and joint activities. The same principles apply to the potential relationships between the peer-group activities and those at home. It should be noted, however, that the educator has a critical role in directing the formal and informal activities of the peer group.

The following example (from Gray 1994) illustrates how an informal setting at a kindergarten created a space for an activity with a group of peers. The group focused on a written label (a name) which had come from one of the children's home. The group played using a Performance tutorial centred on this label, which probably imitated joint activities that each of the girls had experienced at the kindergarten and at home.

A group of five girls (aged between 4 and 5 years) were sitting eating morning snacks. Monica held up her lunchbox which had her name printed on it in black letters.

MONICA	Can you read my name? Look it's M. [points to that letter]
GROUP	[looking at Monica make the sound in unison] mmmmm
MONICA	[looking at each of her friends points to the next letter] a
GROUP	[looking at Monica then one another] aaaaa
MONICA	Now it's n. [puts her finger on that letter and traces it]
GROUP	nnnnn [one friend traces an /n/ with her finger, partly hidden in the fold of her skirt. Another puts her finger out towards Monica's lunchbox, tracing in the air the /n/ shape].
MONICA	Here is i. [points to the letter, traces it]
GROUP	iiiii
MONICA	c [looks at the letter /c/]
GROUP	Ceeee.
MONICA	a is the last.
GROUP	aaaaa.
MONICA	Monica. Monica. [running finger along the bottom of her name]
GROUP	[laugh, looking at one another. One friend touches the letters and murmurs 'Monica'. The others also murmur 'Monica' to themselves. They all laugh again].

A second example from the same kindergarten shows a more direct role of the educator, creating a bridge between an activity at home and the preschool. The role of the peer group is relatively indirect but none the less significant. It is to provide feedback and adopt the role of an audience for a piece of writing. Ginny had made a book with her mother at home. It was created first as an oral text which her mother had then written. The book was slotted into a speech event in the kindergarten — the morning talk (sharing time).

Ginny and the teacher were facing the rest of the group.

TEACHER	Would you like to hear Ginny's story?
CHILDREN	Yes yes.
TEACHER	All right. Please sit down.
TEACHER	Here is Ginny's story.
TEACHER	The story begins — [reads] *This is a friend's house and my house ... I always want to be a witch and I want to be in a witch house.*
GINNY	Yes.
TEACHER	And who are these? [points to drawing]
GINNY	Yes. Yes, this is my family. And this is the grass.

TEACHER	Did you tell your mother what each of the things were?
GINNY	Yes and she wrote that. [points to drawings]
TEACHER	Thank you, Ginny.
GINNY	The end.
TEACHER	Thank you, Ginny. Thank you, Ginny. That was very nice.
CHILDREN	[all clap spontaneously].

Cases of Connecting

An ideal case

An ideal example of the role of concepts and actions in connecting settings comes from a Te Kohanga Reo setting. It does not involve written literacy, rather, a form of cultural literacy (the use of cultural signs and symbols — see Smith 1993). Nevertheless, the principles discussed above were operative here and it illustrates connections in a mesosystem which have developed without a major explicit programme to make connections.

Cazden's (1993b) description of a continuum of teaching can be applied to interventions at the level of systems, too. In this example the programming for connections was one of immersion rather than one of direct and explicit programming.

Te Kohanga Reo can be an extremely effective setting for teaching/learning Maori as a first language. The developmental power of the preschool is linked to the close role of each child's immediate family, even in circumstances where the members of the family are not first speakers, nor perhaps competent speakers, of Maori.

A case study of one 4-year-old's development reveals how co-construction processes in the two settings (Te Kohanga Reo and home) contributed to the developmental power of both settings (Royal-Tangaere & McNaughton 1994). Particular activities co-occurring in the two settings were analysed. At Te Kohanga Reo the activity of singing was a clearly defined speech event which expressed and constructed important cultural messages. For example, it was a vehicle for supporting the formal 'ritual' of mihi (greetings); and it provided occasions for learning about tribal family meanings.

The 4-year-old also came to sing these songs in family settings, a favourite place being in the car going between home and Kohanga Reo. She often initiated episodes of singing, which meant that she was transferring language patterns being learned in one setting (Te Kohanga Reo), in the supportive framework of well-known and meaningful structures, to the second (home) setting, where they were practised and developed further. Within this activity she assumed a role of teacher (or leader — tuakana; see also pp. 32 & 116) for the other family members. She not only taught the song patterns but also used her language expertise in Maori to define and elaborate on words in the songs to her family.

The transfer was due not only to her active reconstruction of the activity of singing. It occurred also because the activity of singing had a presence in the family. Her developing competence slotted into earlier singing activities which were a familiar part of the family's practices since before she was born. The earlier forms of singing were appropriated and transformed, and in certain respects singing now came to look more like the way it occurred at Te Kohanga Reo. For example, the 4-year-old helped to teach her younger sister to appropriate songs, and provided fuller models for incompletely learned songs, mirroring the roles of tuakana-teina present in Te Kohanga Reo.

The incorporation of singing from Te Kohanga Reo into the activity in family settings occurred because the other members of the family were prepared to recognise its significance. They responded contingently to initiations and began episodes of singing themselves. The preparedness of the family to reconstruct the existing activity, and their responsiveness, each contributed to supporting and developing the child's active participation in the event. These processes also took place with other language activities. For example, equivalents of the mihi patterns learned at Te Kohanga Reo developed further at home.

The parents in the family knew the significances of singing in Te Kohanga Reo. They were very familiar with the tutorial systems for teaching/learning waiata. Their familiarity came from their participation in the wider whanau (family or community group) which managed the preschool.

The case study illustrates a well-matched setting. It illustrates the co-construction of expertise within and across settings. Few instances of formal communication, deliberate parent education, and direct negotiation of appropriate pedagogies were needed in this mesosystem.

Direct programmes

> More direct and explicit creation of connections may be needed between home and preschool or school settings, particularly where long-standing practices have not enabled positive connections to be established.

More direct and explicit creation of connections may be needed between home and preschool or school settings, particularly in circumstances where long-standing practices have not enabled positive connections to be established. Examples of interventions for storybook reading at an individual and a group level were provided at the end of Chapter 6 (pp. 118–120).

In these cases the intervention needs to influence features of expertise for parents and educators. These features, of ideas and strategic action which foster connections, need to be developed. Information is critical to this. But critical too are the more subtle variables of who controls the programme and what the goals and benefits might be. This is why at the end of Chapter 6 I stressed the concept of collaborative programmes which aimed to increase access by promoting dexterity.[4]

Summary

This chapter has provided concepts and explanations for how literacy development is influenced by relationships between settings in which children learn. The discussion has been based on the last of the propositions described from Chapter 1:

Proposition Seven: Development is enhanced by the degree to which settings are well co-ordinated in terms of practices, activities, and systems of learning and development. This in turn depends on a number of boundary conditions, including how participants in settings relate to one another.

In the next, final, chapter I turn to more specific descriptions of connections between home and educational settings, and some lessons are drawn for educational practices.

Implications

For families, educators, and researchers

Families: Making connections between settings involves a collaboration in which families and educators relate to, and learn from, each other. On the family side it is important to know about activities and systems of learning and development in educational settings. It is important also to talk with educators about activities at home and how each setting can be responsive to the other. Being responsive includes building on what a child brings home and integrating their new forms of expertise into family activities.

Educators: The implications for educators parallel those for families. Making connections includes knowing about activities and the systems of learning and development in the family. It includes talking with family members, being responsive to their expertise, and relating to and learning from their practices of literacy, so that more effective bridges can be constructed for children's learning.

Researchers: The need for educators to know about and be able to connect with family activities provides some important research directions. Ways are needed of gaining profiles of children's expertise within familiar activities. Models for collaborating with families and pedagogical strategies engaging with and building on children's expertise are needed also. Theoretical issues include plotting the social and cultural bases for effective pedagogy, and determining generalisable principles of pedagogy which apply to local schools and communities.

> It is important to know about activities and systems of learning and development in educational settings.
>
> Making connections includes knowing about activities and systems of learning and development in the family.

Further Reading

Matched settings:
Cazden, C. (1988). *Classroom Discourse*. Heinemann, Portsmouth, NH.
Cummins, J.(1986). 'Empowering minority students: a framework for intervention.' *Harvard Educational Review*, 56, 18-36.
Tharp, R. G. & Gallimore, R. (1988). *Rousing Minds To Life: Teaching, Learning and Schooling in Social Context*. Cambridge University Press, Cambridge.

End of chapter notes

1. Given that it does matter what families do (e.g. in the case of storybook reading for narrative meanings), how should we interpret differences? For example, if some groups of families do not read storybooks employing the narrative activity, or if some families infrequently focus on and elaborate meanings, *and* this is related to achievement at school, how should their practices be considered? Typically there are two approaches to answering this. The first is to assume that their activity is inadequate and therefore the parenting is deficient. A second approach argues against this and for recognising and celebrating the diversity of social and cultural and social forms. This book adopts a modified form of the latter view. Activities do have social and cultural bases. But some forms of literacy have differential consequences for access to school forms of literacy. The view adopted in this book is one which values dexterous development and access to those forms of literacy that are essential to educational settings (see McNaughton 1994b).

2. Dyson (1991) has described the dynamic personal constructions of children in a first-year classroom as they made connections between forms of writing with which they were familiar and those promoted in the classroom. What is remarkable about Dyson's close observations is the degree to which the children worked hard (and enjoyed) seeing connections and tried to make the writing activities at school meaningful in terms of their knowledge of forms and content outside of school. But it is also clear from Dyson's analysis that the role of the teacher was critical. The teacher created a Zone of Promoted Action which was responsive to the children's initiations and their playing and experimenting with forms.

3. The report by Thackery *et al.* (1992) provides details of the results of the survey and interviewing in 100 schools. The report notes that: *It is customary for schools to find out whether children have shown an interest in writing, can write their own names, and, perhaps, can write some other words from their personal vocabularies*, although this may be collected after some days at school (p. 44). More specifically, their results revealed that about one third of the schools asked the child to provide a sample of writing using pencil and a blank sheet of paper. About one quarter of the schools established whether there were any words the child could write freely. Twelve schools assessed ability to form letters.

4. Heath's (1987) intervention for early literacy and language at a community and classroom level represents an effective form of direct programming. As noted in Chapter 6, several elements of effective intervention were present. The intervention included procedures both for developing new ideas and for acquiring strategies for action. But the researcher acted to 'connect' families and teachers directly, collaborating with them to develop these ideas and strategies.

Chapter Ten

Resourcing families and educators

> **Focus**
> **Effective educational provisions for families**
> - Families need to have access to ideas and strategies that are associated with educational success.
> - Educators have a critical role in developing effective educational practices for children and their families. The general challenge is to develop shared goals and activities with families.
> - This challenge requires educators to develop particular ideas and strategies about the nature of learning and development at home.

The previous chapter provides a framework for analysing literacy development in family, educational, and other socialisation settings. It can be used as a means for making recommendations about effective educational provisions.

This final chapter takes the ideas and evidence of the previous chapters and uses them to discuss how expertise can be fostered in family settings. It looks at the role of early-childhood and school settings in this nurturance. This leads to recommendations about family and educational practices. In a sense this has been the most difficult chapter to write because the central argument of the book has been that what develops reflects and constructs the values of the participants. If recommendations are made, they have to be made with the qualification that they reflect the values that I hold.

The conclusions and recommendations that follow are guided by three assumptions. The first is that diverse forms of literacy in sociocultural groups should be valued. The second is that I assume universal processes operating within these forms of literacy can and should be identified. This means locating those processes that are present in, and foster development across, diverse literacy activities. The concept of effective tutorial configurations is one such example of the search for universal processes.

The third assumption acknowledges that our schools have the task of fostering particular forms of literacy — those that are valuable and effective for the tasks of learning at school, and for work and living outside of schools. But families have a right of access to those forms of literacy which are recognised at school. This means they have a right to know about and have resources to practise these forms at home. In turn, schools have the challenge of finding ways of bridging literacies.

> Schools have the task of fostering particular forms of literacy, but families have a right of access to these forms, and to know about and have resources to practise these forms at home.

What is known about relationships between settings? Perhaps more significantly, in which areas might practices be less than effective and able to be improved as judged against the assumptions I have just outlined?[1]

Mesosytems before School

Preschool

Bronfenbrenner's proposition about mesosystems predicts that the relationships between developmental settings will have an important effect on development in each setting. One of the significant transitions and ongoing connections for children before school occurs when they enter a preschool (early-childhood educational) setting. The potential for significant relationships affecting literacy would be present across these settings.[2]

A range of activities occurs in preschool settings. Some of these could complement those occurring at home, creating the wide significance for children I discussed in the previous chapter which affects the further development of Personal and Tutorial systems.

> A range of activities occurs in preschool settings, some of which could complement those occurring at home, creating a wider significance for children.

Most preschools have an area where books are available either for specially organised book activities or for when choice of activities operates. This space is used regularly in many preschools to foster book-reading activities, either directly through tutorials with individuals or groups (e.g. in shared book reading), or indirectly by making the space salient and inviting, and by responding to children's interest and engagement with texts.[3]

Another relatively frequent activity is the naming of paintings and other work completed at preschool. This standard practice results in children often bringing home named products from preschool. Writing a name on a painting is an ambient activity which children can observe and from which personal constructions can develop. The effect on personal constructions depends partly on how salient this is for the child. And this in turn is dependent on the talk in this activity, as well as the knowledge that the child can bring to bear. Similarly, a tutorial system can develop around this activity also with the properties of writing to name described in Chapter 7.

But whether or not these activities are fostered directly (through tutorials) or indirectly (by providing conditions which support personal systems) is dependent on teachers' beliefs and attitudes. In New Zealand the general developmental model guiding early-childhood education has come to be constructivist

(Nash 1991; Podmore & Bird 1991. See also Chapter 1). It is not clear how this general theoretical framework might influence curriculum arrangements for literacy. But there are indications that only a small number of early-childhood educators believe that early-childhood education centres should be involved in formal emergent literacy programmes (Foote & Regget 1992).

> There are indications that only a small number of early-childhood educators believe that early-childhood education centres should be involved in formal emergent literacy programmes.

This is what Tizard and her colleagues found when they asked 36 teachers in nursery classes in London which skills they expected children to acquire by the end of nursery school (Tizard *et al.* 1988). Only about half of the teachers expected children to write their name. More than three-quarters of the teachers, however, did expect the children to know that we read print, not pictures, and that we read from left to right. But only a quarter expected that the children would acquire knowledge of letters. Tizard describes these teachers as differing markedly in the emphasis they gave to literacy activities, and in general as having limited expectations.

When the teachers were asked how they tried to ensure that children learned some of these things, they generally referred to children being immersed in a print environment. Teachers generally did not refer to specific teaching. That is, their expectations were closely related to what they attempted to promote. One exception was in providing models for names which children copied. Some teachers mentioned that official policy suggested they were not supposed to influence children directly.

It is clear that specific beliefs are influential. Catherine Snow, David Dickinson, and their colleagues (Snow 1991; Dickinson *et al.* 1992) have provided detailed descriptions of the literacy and language qualities of preschool settings for a group of low-income children in the United States. Their analyses indicate significant differences between preschools in the literacy activities in which children engage. They have shown that the pedagogical beliefs of the preschool teachers were closely linked to the activities children experienced. That is, their beliefs determined the Zones of Promoted Actions (see p. 18) that operated in their preschools.

The most striking differences were between teachers whose attitudes reflected an orientation to support language and literacy development (through, for example, book reading in groups and print-associated talk) and teachers espousing a *laissez-faire* approach, in which large amounts of free time were available and the emphasis was on social and emotional development. In the former teachers' classrooms children were engaged in significantly larger amounts of talk relating to books and narratives and print conventions.

These researchers have drawn implications from their studies for preschool educational programmes. They make specific suggestions in three general areas. Firstly, they point out the need to reduce time spent organising children, which was very high in some preschools. Suggestions included smaller groups and arranging an effective physical space for reading. They then make detailed suggestions for how to encourage elaborated talk during book reading, and how to reduce teacher dominance during discussions. Each of their suggestions is applicable to the context of preschool education in New Zealand. But, in addition, a theoretical framework which provides a rationale for organising and deploying resources for literacy learning in preschool settings is needed, too. Being more deliberate in promoting literacy activities does not mean necessarily being more direct. It does mean arranging, selecting, and deploying resources for both the early-childhood and home settings.

> Being more deliberate in promoting literacy activities does not mean necessarily being more direct.

Relationships between preschools and families can be fostered in other ways. Several mechanisms can be illustrated from the SOL study. Direct linkages occurred in several ways. One level was the involvement of parents in the early-childhood setting. Five of the Pakeha mothers (and the father in a sixth household) had served, or were serving on playcentre, kindergarten, and school committees. One of the Samoan mothers had such a responsibility, too. These links enabled activities in one setting to be known about and actively transferred to or recognised in the other (home).

Clearly, creating these links depends on parents resources, their beliefs about appropriateness, and their strategies for involvement. But this in turn is dependent on what the preschool settings can offer and how conducive they are to supporting involvement from parents who may have different communication styles, different meeting styles, and different decision-making styles from those typically associated with committee work.

> Creating links between preschools and families depends on parents resources, their beliefs about appropriateness, and their strategies for involvement. But this depends on preschools supporting parental involvement.

Direct communication between early childhood education settings, families and schools is another obvious component of connections in the mesosystem. Given that preschools do actively promote literacy activities it is essential that profiles of children's developing expertise follow children to school.

Other settings

Another relationship between systems in the SOL study was very strong, especially for the Samoan and Maori families. It was the relationship between the church (including Sunday school) and the family. Two of the Maori mothers and five Samoan parents held positions at their respective churches which involved them in duties that included activities in the Sunday school.

The same outcomes are applicable here in terms of the general developmental power of both settings for children. The degree to which the activities of home and Sunday school were mutually complementary could be expected to have a significant impact on the developing systems in both settings. What is important here, however, is that the churches took an active role in promoting particular sorts of literacy activities. The connections between the home and the church, especially at the level of complementary literacy activities, were very strong in these families.

This was very obvious in the Samoan families in the study. Each family had an alphabet chart at home, similar to the chart on page 183. The preschoolers had all been taught letters using this chart. This teaching with similar resources happened at church as well as at home. In fact, Samoan churches in general are instrumental in promoting this literacy activity as well as others. In these families there was a close relationship between settings, where the activities were similar, and where the pedagogies associated with the activities were similar. This close relationship enabled the developing systems to be mutually reinforcing

Sample 10.1 An alphabet chart used by Samoan families in the SOL study

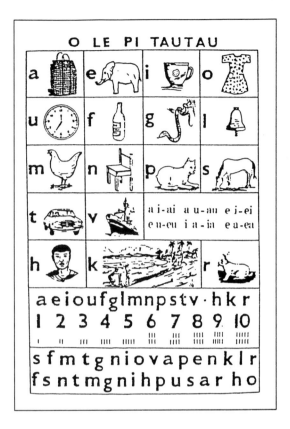

The degree to which other sorts of connections exist for some families between these two settings is illustrated with the development of Tongan children referred to in Chapter 6. The style of reading storybooks in their families is described on page 113. It produced a tutorial system focused on Reading for Performance. The 4-year-olds in that study were also observed at Sunday school. All of them went to the same church, and at Sunday school they were taught hymns, prayers, and articles of faith. Performance tutorials were used in these activities. A sample from a transcript of their Sunday-school class learning a new hymn is shown below.

Sample dialogue 10.1 A segment from a Performance tutorial for learning a hymn in a Tongan Sunday school. (from Wolfgramm 1991)

TEACHER	... Taha ... Hiva!
	(... One ... Sing!) [pointing at the board]
CLASS	[sings the first line of the hymn]
TEACHER	Toe ai ... Ua ... Hiva!
	(Again ... Two ... Sing!)
CLASS	[repeats the first line of the hymn]
TEACHER	Sai ...
	(Good ...)
	[sings second line of the hymn]
TEACHER	A'i pe ia ...
	(Sing just that line)
CLASS	[sings the second line of the hymn]

Placed alongside examples of Reading for Performance at home, the transcript demonstrates the presence of shared activities, shared pedagogies, and shared literacies. Again, there is evidence of mutually reinforcing settings where knowledge and developing systems are shared. More is shared than the form of teaching and even what is taught. Values, beliefs, and expectations are shared. The two systems can be said to be linked in a cultural and social endeavour.

The implications of this analysis for educational practices in preschools and schools are twofold. One is the need to appreciate the significance of this setting, and the relationships it has with the family, for literacy development. A closely related implication is the need to consider the role of this connection in educational programmes. That is, if programmes are being developed, the role of significant child-rearing settings such as the marae, or church need to be considered, both in terms of collaborative practices and in terms of knowledge upon which to build teaching decisions.

> Preschools and schools need to appreciate the developmental significance of child-rearing settings such as a church.

The Transition to School and Home–School Connections

Another major mesosystem is created when a child makes the transition to school. The connections between the home and school settings also can be explored along several dimensions of familiarity and co-ordination between settings.

One aspect of familiarity is in the similarity of activities at home and at school. Similarity exists if forms of expertise and ways of teaching and learning are shared. In addition to the familiarity of shared ways of doing things there are similar topics in what one talks about, what one reads about, and what one writes about. This familiarity is achievable when topics and purposes within activities which are peculiar to children are shared with teachers. Topics may be special to children as individuals, as a member of a particular family, and as a member of a particular sociocultural group.

Familiarity and co-ordination between settings are based on the roles of teachers and parents. In particular, they are based on the ideas and strategies that parents and teachers have relating to children, activities, roles, and pedagogies (see Chapter 9). Questions can be asked about the degree of co-ordination between homes and schools and about the conditions that influence effective transitions over the first year, such as the ideas held by parents and teachers about contact. When research attention is brought to bear on these features of connections between homes and schools it reveals some important strengths and some telling weaknesses.

Making contact: the role of ideas

The ways in which parents value and evaluate classrooms (for example, as places where parents can help), and the ways in which teachers view parents (for example, parents' presence in the classroom) provide one set of ideas that determine connections. These ideas come to influence the way parents deploy resources such as their time and energy making contact with classrooms.

When the children in the SOL study were preschoolers all but two of their parents believed that parents should have a lot of contact with the school. That belief in frequent contact continued over the first six months of school. It provided a rationale for actively seeking connections, in that the parents' reasons were in all cases to do with knowledge that parents wanted about how their child was developing at school.

But in several instances this interest was tinged with a concern to check and act on concerns. A Pakeha mother believed frequent contact was important because *'you know how your child is being taught'*, and added, *'Up north mothers took it into their own hands. One mother was designated to put in a complaint'*. In rationalising the significance of this contact several parents claimed that their child's progress was attributable in some way to their involvement: *'I guess a lot of it has to do with my help'* or, *'Children are happy when mother is involved.'*

This rationale translated into action. At each interview the parents estimated how much contact they had over a week. This was defined in terms of being physically present in the school or talking with school personnel, and included all types of contact, such as helping to make lunches at school, telephoning the school secretary, and talking with teachers. When the child was a preschooler (with an older sibling already at school) the parents estimated they had an average of two contacts in a week. At 5 years 6 months the average was still two contacts a week, although compared with the Pakeha families, Maori and Samoan families were more likely to report low or no contact. The strength of this contact was illustrated when eight of the parents spontaneously mentioned they had formed friendships with school personnel.

It must be remembered that these families were selected as having a child already doing well at school. Their sense of familiarity with contacting school can be seen as having laid the foundation, as well as being an accompaniment and a product of this previous experience. There is some evidence that parents with children who are not progressing well become increasingly frustrated and reluctant to forge connections with school. The evidence is not that they are not interested or motivated to foster their children's development (McNaughton, Glynn, & Robinson 1979).[4]

> Parents with children who are not progressing well become increasingly frustrated and reluctant to forge connections with school.

Making contact: seeing expertise

An important implication of discussions in previous chapters has been the need for schools to broaden their view of new-entrant children's expertise. In previous chapters I have documented how narrow the focus has tended to be at this transition point, a focus that is formalised in what information schools do and do not attempt to gather.

The same argument that was made for teachers in early-childhood-education settings applies here, too. What one is prepared to see is dependent on concepts about development and learning. Similarly, a teacher's expertise in looking for children's literacy is dependent on the strategies they have for gathering information. Teachers need to develop procedures for gathering profiles of children's expertise and the activities (and settings) in which these have developed.

> What teachers are prepared to see depends on concepts about development and learning.

Without such procedures it is difficult to develop shared goals and useful tutorial structures that make bridges between expertise out of school and in school.

School activities at home: reading

Connections can be forged by schools in practices that are disarmingly simple and seemingly unremarkable. One such practice is a very important feature of beginning reading in New Zealand schools. It is sending books home for children to read to family members (see Sample 10.2). This has been a recognisable national policy since at least 1962 and can be considered from its habitual and rule-bound nature to be a national literacy practice (see McNaughton et al. 1992b). There are few parallels in other countries to this as a national practice.

Sample 10.2 Examples of family members confirming with a signature on a sheet sent home from school that they had heard the child reading their book from school. These records includes grandparents' and even a great-grandparent's signatures. The signatures represent a considerable commitment in terms of the organisation of time, as well as a significant teaching role

> Sending books home to read deliberately creates a link between home and school.

Sending books home to read is a component of the mesosystem; it deliberately creates a link between home and school. Information can flow via this link so that learning systems in the two settings are more closely attuned, with activities and tutorial systems becoming complementary. An illustration of the significance of this for families is shown in the records kept by one family of having the books sent home. The simple record keeping was part of the school's organisation. Family members (including grandparents and great-grandparents) signed that they had heard the reading.

From this perspective the family participates in an important literacy practice that can be predicted to have significant effects on the development of reading within the activity of oral reading as defined by the school. The prediction comes from Bronfenbrenner's arguments about the nature of mesosystems, and from the increased opportunities to practise across settings and across applications.

Obvious conditions make the family's role in this practice more or less effective. They include knowing about and being able to engage in appropriate tutorial systems. This in turn is partly dependent on the child's role as an expert reader of each book, which in turn is dependent on books being sent home that guarantee high levels of accuracy and self-regulation.

Sending books home has been examined quite closely. It is an almost universal feature of junior classrooms in New Zealand, with both teachers and parents reporting that books go home for reading three or more times a week (McNaughton et al. 1992b). Teachers and parents believe it should occur and both groups express considerable confidence in their role in the practice. The number of times it occurs a week (at least three times) has been shown in experimental studies to be an effective rate to have a positive influence on learning to read at school.[5]

> Teachers sending books home at least three times each week has been shown to be an effective rate to have a positive influence on learning to read at school.

Families and schools could be more closely linked, however. One indication of this comes from research evidence of a mismatch between parents and teachers about whether or not parents have gained the knowledge and strategies for what to do when hearing their children read (McNaughton et al. 1992b). In that research teachers believed that they had told parents what to do, and this information was typically noted on the folders in which children take their books home. And some schools did, in fact, provide quite specific guidelines (e.g. Sample 10.3, p. 188). Few parents, however, identify school-initiated information as the resource they use for trying to help.

In one study, 49 parents of first-year children were interviewed (McNaughton et al. 1992b). Half of these parents said they wanted more information. On the other hand, when the 18 teachers of these parents' children were asked whether they knew what the parents did, they reported that they did not know anything about what the majority of the parents actually did when they heard their children read.

In such an information vacuum, privileged information about school activities that some parents might have increases in value. Many of the parents in the study who rated themselves as very confident in helping their children referred to access to (prior) knowledge about teaching practices. These included personal

Sample 10.3 An example of instructions sent home with a child for family members about how to hear the child reading schoolbooks which are sent home

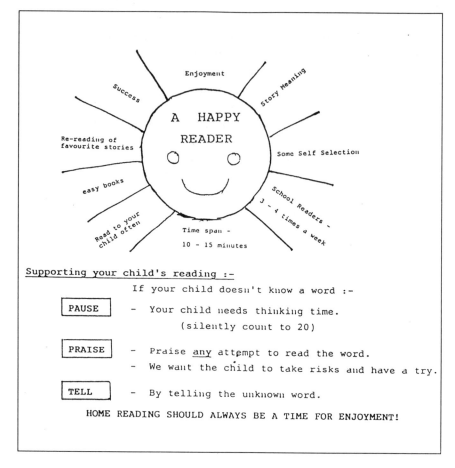

> In the absence of information parents develop their own theories in practice, which may or may not match those of the school.

resources; for example, a parent would point out that they had direct experience in the educational system. They would comment that they were an *'ex-teacher and remedial teacher'*, or had *'a degree in Education with a paper on reading'*. Resources were also located in the wider family; for example, some parents said that they had an *'uncle a teacher'*, or *'four in the family are teachers'*.

In contrast, other parents said they taught themselves through observation and trial and error; they would *'learn with (the) child with material that he has brought home'*, or *'by listening, generally answering* [the] *child's questions'*.

In the absence of information parents develop their own theories in practice, which may or may not match the theories of the school. It has been found that parents who did not have close contact usually interact in ways with their children that are different from the typical classroom routines in oral-reading interaction.[6]

One of the conditions identified in Chapter 9 for a more integrated mesosystem is how the parent's role is conceived, both by themselves and by teachers. Table 10.1 (p. 189) shows the responses of 18 junior-class teachers and 49 parents of the children in their classes to the question about what the role of the parent should be in this practice of sending books home. The majority of both parents and teachers said that the role should be supportive of the

Table 10.1 Beliefs of parents and teachers regarding parental role in hearing reading. (Percentages of parents or teachers)*

Role	Parents	Teachers
Teach	34.1	0
Support	56.8	92.1
Other rules	9.1	7.9

*Parent reports come from 49 interviews; teacher reports are based on interviews with 18 teachers, each referring to three specific parents.

(from McNaughton et al. 1992b)

child's learning at school (for example, to make sure reading is done and ensure practice), but should not be seen as an active teaching role.

This is not a surprising finding. For many years New Zealand governmental publications have referred specifically to the parent's role as supportive. The most recent publication (*Reading in Junior Classes*, Department of Education 1985) refers to books being sent home for parents to read with their children as part of a *close working relationship* [*which*] *is essential* and that *the senior teacher will be involved in ... explaining to parents how they can help their children at home and support the school programme* (pp. 148–149). It is interesting that while none of the teachers interviewed specified a teaching role, a significant number of the parents (15) said they should have a teaching role.

This way of viewing the role of the parent tends to reduce teachers' perceived need for systematically sharing information about instruction and about reading practices. This view is not an accurate description of what parents are in fact asked to do and what they actually do, as shown in the transcript from the SOL study (p. 190).

The transcript is a record of a mother hearing her daughter read two books she had brought home from school. One was more familiar and easier than the other. At strategic points the parent praised (*'Good girl'*). When the child had halted, she paused rather than prompted, which provided an opportunity for the child to self-correct (*'I can read to my ... '* [pause]), and she provided feedback (*'No, it only says ... '*). She also prompted for error correction, using a gap-filling routine (*'Peter ... '* [voice raised on last syllable, as a query]).

All of these interactions are instances of teaching acts. It is important to note that they were able to occur in the context of a relatively easy text. The choice of an easy text meant that the child's expertise constrained what the mother did. The good performance influenced her good teaching practice. But when the daughter had a book that she found more difficult, the mother resorted to performance routines (*'Are in bed. ... Are in bed'*). Performance routines are not intrinsically a problem. But in the context of a task in which the reader is meant to be practising effective comprehension-based strategies, reliance on them is not appropriate. However, it would have required more expertise of the sort applied in classroom instruction to scaffold this book in a manner which would have increased the level of accuracy.

Sample dialogue 10.2
Mother hearing child (5 years 1 month) reading school books. (SOL study, McNaughton & Ka'ai 1990)

Text 1 (easier)

Child	I can read to my ... [pause]
	I can read ... [pause]
	I can read to my Mummy.
	I can read to my Daddy.
Mother	No. It only says 'I can read to Dad'. It doesn't say 'my'.
Child	I can read to Grandpa.
Mother	Nana.
Child	Nana.
	I can read to sister ... [pause]
	I can read to my sister.
Mother	That's better.
Child	I can read to my ... [pause]
	I can read to my cousin.
Mother	Friend.
Child	Friend.
	I can read to myself.
Mother	Good girl.

Text 2 (more difficult)

Child	Peter and Sally are sleeping.
Mother	Are in bed.
Child	Are in bed.... [pause]
Mother	Peter ... ? [voice raised on last syllable, as a query]
Child	Peter shouted and shouted. Sshh ... [pause]
Mother	No. You are naughty to shout.
Child	You are naughty to shout.
Mother	Said Sally.
Child	Said Sally.
	Sshh. Mummy and Daddy are asleep.
Mother	Mother and Father ... [voice raised on last syllable, as a query]
Child	Mother and Father are asleep.
Mother	Good girl ...
	I am naughty.
Child	I am naughty.
Mother	I am naughty ... [voice raised on last syllable, as a query]
Child	Shouted Peter.
	Mother said.
Mother	You are naughty, Peter.
Child	You are naughty, Peter.
Mother	Said Mother.
Child	Said Mother.

By constraining the parental role to one of support the relationship between parents and teachers is defined in terms of the teacher having educational power. This maintains any uninformed pedagogical practices at home as inadequate or inferior. The problem with this form of relationship is that the participants do

> By constraining the parental role to one of support, the relationship between parents and teachers is defined in terms of the teacher having educational power, and there is a reduced need to learn from each other.

not have a rationale for, nor an opportunity to learn from, each other. Beliefs about relationships provide a condition for effective linkages between microsystems. It would be more conducive to connecting the systems if the role was conceived as adjunct teaching.

There are other less direct ways in which learning to read at school may influence family activities. For example, some parents still read to their children after their child goes to school.

In the SOL study all but one family said they continued to hear the child reading books brought home from school. Parents had different rationales for the continuation of this activity. What seemed to be common was a sense of mutual enjoyment. Parents said they were *'glad it happens'*, *'it is a time for us to share together'*. The corollary of this, a sense of something lost, comes from one mother who said her eldest son still read to her daughter but her daughter insisted on reading to her. *'She doesn't enjoy being read to. She's very independent now.'* This oral book-reading activity, precious to some families, may be inadvertently interfered with by the transition to school and activities which stress a notion of solitary and child-controlled reading.

School activities at home: writing

It is not common for New Zealand teachers in junior classes to parallel in writing what they do in reading. They do not deliberately set up writing activities to be carried out at home. Yet records from a number of studies clearly show that such writing, including equivalents of caption books and constructed stories, is done at home. In the SOL study nine of the mothers mentioned such a practice at home and examples were collected, including an unsolicited tape recording on which this activity had been recorded.

What is so interesting about the writing examples at home is that they show two aspects of generalising activities across settings in the mesosystem. The first is the active agency of the child in transferring these activities (and the expertise) through personal systems. The second is the active agency of parents in responding to the initiatives of the children and contributing to the development of tutorial systems and therefore personal systems for these writing activities at home.

A sample of the transcript from an audio recording is shown on page 192, along with part of the completed writing by the child, aged 5 years 6 months. The mother was busy preparing tea at the time the tape was recording, and the child was close by trying to write an account of an event. The mother responded to her child's questions and problems, and anticipated text, helping the child to form a clear account. She was able to do this of course because they shared the topic and important conditions for the development of complementary activities discussed in Chapter 9.

The purposes and exact meanings of parts of what was said are difficult to interpret. This was a fragmented and interrupted conversation. But clearly, an activity system for writing narratives occurred here, which involved both collaborative participation and embedded routines for spelling. The child signals the need for help ('… *somethink*'; '*Mum, what's this?*') and the mother provides guidance for obtaining appropriate oral accounts and for spelling. The tutorial system included a phoneme–grapheme (sound–letter) segmentation strategy (p, … p… i-n-g, etc.) and specific highlighting of print conventions ('*Full stop*').

Sample dialogue 10.3
A transcript of a mother collaborating with her child (5 years 6 months) in writing a narrative in the form of a personal account. The Tutorial occurred at home and was initiated by the child. Embedded within the collaborative participation were performance and labelling routines.
(SOL study, McNaughton & Ka'ai 1990)

M…	I went to the shopping somethink? I went … [pause]
MOTHER	That's when, eh?
M…	To? went to … [pause] the … N … N. Mum, what's this?
MOTHER	h, o
M…	o
MOTHER	p
M…	p, p … i–n–g … shopping … i–n–g is it … [pause; tape unclear — child pronouncing shopping]
MOTHER	What is it?
M…	ping? … T
MOTHER	You know.
M…	H?
MOTHER	and here?
M…	T, g … um … g …
MOTHER	No, n, n … not h.
M…	I went …
MOTHER	No.
M…	I went to the shopping centre then we [pause] we went …
MOTHER	Home [with emphasis] home. You know we have to go somewhere.… went home.
M…	I … oh …
MOTHER	Full stop.

Sample 10.4 Part of the child's personal narrative completed during the collaborative dialogue above

> I went to the shopping. Then we went home
> We went to t
> I went to the Beach
> to Swim Because
> I was hot.

> I went to my cousins
> We had a party,
> & then we went to
> Play. It was
> dark so we went
> home.

> Educators need to support complementary activities at home, and develop shared goals and activities with parents.

In general, there is a need for educators to be clearer about supporting complementary activities at home. This can be viewed as an educational challenge, to develop shared goals and activities with parents. The potential for shared practices is illustrated in the practice of sending books home and in the personal exploration and practice of writing activities that children transfer to home.

A Note on Exosystems

The general focus in this chapter has been on the properties of what Bronfenbrenner called the mesosystem (see p. 164). In other chapters I have discussed how culturally based ideas and values influence developmental processes. These ideas and values correspond in part to the overarching system which Bronfenbrenner termed the macrosystem. In between the macrosystem and the mesosystem he posited another system, the exosystem (see p. 164). It includes those parts of a family's life that the child may not be involved in directly but which nevertheless impinge on a family's child-rearing practices. These include the resources in the neighbourhood and the world of work. I discussed some aspects of how these impinge on literacy practices in Chapter 2, but a further comment on the properties of this system as they affect literacy activities is appropriate here.

> The exosystem includes those parts of a family's life that the child may not be involved in directly but which impinge on a family's child-rearing practices.

Exosystems are influential and dynamic. In Chapter 2, families were described as resourceful, meaning that families use resources in carrying out their child-rearing tasks. In turn these resources are dependent on a multitude of circumstances, such as living conditions, working conditions, income levels, networks of support, and access to health, education, and welfare services. An analysis of the family and its relationship with educational systems needs to include an examination of these circumstances. For example, the analysis is incomplete if it is assumed that all families live in circumstances that mean they can socialise their children effectively, meeting the goals they have set. A discussion of mesosystems also contains implications for educational policy including the need to support the robustness or resilience of families.[7]

> Analysis of the family and its relationship with educational systems needs to include an examination of the family's circumstances.

Events take place and family circumstances change which impinge on family functioning. For example, the degree to which a family can contact the school, or the teacher has easy contact with a family is dependent on a number of circumstances. Earlier, in discussing literacy activities (p. 21), I used the example of the stressful circumstances one family was living under which stimulated the activity of writing letters to social agencies in an attempt to alleviate these circumstances. The stresses on this family continued before and over the transition to school. Just before the young child was to start school the family shifted, and the older sister had to attend her fourth school in three years. The family reported that the nearest school told them that the school roll was closed. The next nearest school was several kilometres away.

At 6 years of age, the child we had been following over the transition to school told us that she and her sister walked home from school. Both her father and mother were unemployed (outside the home). The school knew very little about the family. There was little shared familiarity. Difficult accommodation, repeated shifts, unemployment, lack of transport, and perceived rejection from the local school, to name those variables of which we were aware, may have conspired to make it difficult for the family to keep in touch with the school. The younger child was one of two in the SOL study, who, despite predictions, made (relatively) low progress at the end of the first year of school.

Another child who was predicted to make high progress made erratic rather than low progress. Her profile is represented in Figure 10.1 (below), and reflects elements of the exosystem influencing the mesosystem and ultimately the functioning of microsystems in which she was involved. Her progress before school and over the transition is shown against the backdrop of the average profile for the other children in the SOL study. It needs to be remembered that the average scores in Figure 10.1 are for one narrow slice of the expertise that these children were developing. It is the sum of three subtests on Clay's (1979) diagnostic survey — concepts about print, letter identification, and words written.

As can be seen in the graph this child was slightly below the group average in this knowledge before school and slightly above the average in the first months at school. During this time she and her family shared the common characteristics of the other families. These included the presence of extensive home-literacy activities and close familiar relationships with the local school.

Two major life events took place half way through the year. Her mother and father separated and she and her sister stayed living with their mother. As a consequence she changed school. From this time there was a problem with the relationship with the new school.

When the child was 5 years 6 months old, her mother said she had been too busy to do some of the things she ordinarily did, such as reading stories with her daughter. At 6 years of age, her daughter was reported to be happy although lacking in confidence at the new school. She did not often take her reading book home. Both the teacher and her mother had begun to realise this but neither was in direct contact with the other and did not know what was happening and what the expectations might be in the other setting.

This girl's profile shows how development and child rearing are dynamic. Events in the exosystem influenced the mesosystem and the functioning of activities at home. The reduced rate of acquisition between 5 years 6 months and 6 years was associated with major changes. In keeping with the earlier description of processes of generalisation (see Chapter 9, p. 167) the child was active in develop-

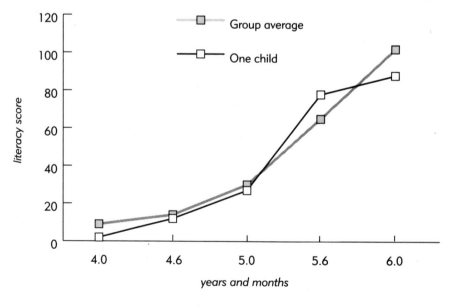

Figure 10.1 A child's concepts about print, letter identification, and writing, over the transition to school. The child's development slowed between the age of 5 years 6 months and 6 years, in comparison with earlier development, and in comparison with the rest of the group (first) shown in Figure 8.1 (p. 149).
(from McNaughton & Ka'ai 1990)

ing the nature of the relationships between home and school. In some respects this child may have actively contributed to trying to alleviate difficulties at home. Her solution to the immediate problems of extra pressure at home was to *not* take books home for her mother to hear her read.[8]

> Provisions for literacy activities in the home cannot be separated from the life experiences of a family. Whether or not a family reads storybooks to their children is dependent on their psychological and social resources.

Each of these examples is used to underline the point that provisions for literacy activities in the home cannot be separated from the life experiences of a family. Whether or not a family reads storybooks to their 4-year-old, or hears their 5-year-old read school books effectively is dependent on their psychological and social resources. The success of educational practices is dependent on social and political policies which affect the family. Children's literacy development is dependent on neighbourhood resources and local and national policies which affect family life and functioning. Of considerable importance are those policies which enable families to spend time in literacy activities and provide the resources needed for those activities.

Summary

The elements of the socialisation model (Chapter 1) provide a tool for explaining how and why learning and development take place. In this last chapter I have explored areas in which teaching practices in early-childhood and school settings can be made more effective. I have emphasised two components of teachers' expertise which contribute to making effective connections between settings. One concerns their ideas and values about pedagogy, learning, and development, and about their professional roles. The second component is the strategies that teachers can develop for applying their ideas.

Implications: Lessons and Challenges

For families, educators, and researchers

This book has attempted to review and contribute to ideas and strategies in emergent literacy — for parents, educators, and researchers. My concern has been to contribute to the provision of educational programmes which promote expertise both in home and school forms of literacy, for children from diverse social and cultural groups.

> Promotion is needed of expertise in home and school forms of literacy, for children from diverse social and cultural groups.

I have described the early development of literacy as a co-construction. Literacy activities are threaded through child-rearing environments. Systems of learning and development take form in these activities, and through them children develop expertise in literacy. The activities are part of a family's practices, their ways of being in the world. They reflect and construct social and cultural meanings. They are shaped by and in turn shape the environments within which the family functions.

The challenge now for teachers is to develop strategies for fostering direct and indirect linkages between settings. This will include having activities that are shared or complementary. It will include sharing ideas and values, as well as establishing practices which support the generalisation of expertise across settings.

This challenges teachers and researchers to design procedures for gaining profiles of children's expertise which are sourced to activities at home. Profiles are needed while children are in early-childhood education and at the point of transition to school.

> Profiles are needed while children are in early-childhood education and at the point of transition to school.

Another challenge exists in the curriculum and in teaching practices of early-childhood education. In early-childhood settings (before formal instruction) immersion in literacy activities provides a basis for children's development through their exploration and experimentation. But there is also a need to select, arrange, and deploy activities carefully within those settings. The development of effective joint activities between teachers and children and the development of effective peer-group activities are important concerns.

There is a need for parents to gain access to ideas and strategies that are significant for educational success. This provides a basis for informed choices about child rearing. Teachers have a critical role to play in developing effective educational practices for children and their families. As educational researchers, we have critical roles also. The research agenda in emergent literacy includes unravelling the complexity of transitions, understanding how teachers can construct bridges to children's expertise, and underpinning studies of literacy with fully informed analyses of cultural and social processes.

> Parents need access to ideas and strategies that are significant for educational success.

> Research needs to understand transitions, literacy bridges, and cultural and social processes.

End of chapter notes

1. It is impossible to predict completely the developmental outcomes for a particular child. Partly this is because children are active agents in their own development and this leads to a partial indeterminacy in predicting development. But our inadequacy turns out to be a general problem of explanation in sciences which look at development. The problem is expressed best by the evolutionary biologist Stephen Gould in his book *Wonderful Life* (Gould 1989). He argues that a central principle in all 'historical' explanations is the presence of contingency, that is, there is an inherent unpredictability about explaining development in living things across time. Unpredictability occurs because unanticipated events occur, experiences take place that are unconnected with the developing organism.

Gould points out that we can offer a full explanation in retrospect for why the dinosaurs died. But we could not have predicted their demise. Similarly, we cannot predict with a single child the detailed specifics of their literacy development. We cannot say with certainty for a particular child that even when particular activities are promoted that this will inevitably lead to specific developmental outcomes. Life is just too contingent (on other events that occur in the lives of families).

However, the theoretical framework developed in this book means we can talk about the mechanisms that co-construct development. We can develop a *degree* of predictability in our explanations. We can assume that the probability of certain outcomes taking place is very high in the presence of certain activities and, afterwards, that particular events occurred which enabled development to occur.

Knowing with a degree of certainty that these events could occur allows us to predict with some sureness that learning will occur and how development will take place.

For example, given that reading storybooks to children (within the activity of reading for narratives) enables expertise to develop which gives effective access to other forms of literacy at school, then certain recommendations and predictions can be made. Families need access to particular beliefs, values and knowledge, and the resources which support these. This is the challenge for educators concerned with access to education and achievement for children from different social and cultural groups.

2. It is estimated that in New Zealand above 90 per cent of all children who are between 4 and 5 years old go to some form of preschool. The figure is above 95 per cent for non-Maori and about 75 per cent for Maori. These figures drop to about 75 per cent and 60 per cent respectively for 3-year-olds (Ministry of Education 1993). These settings, therefore, represent a significant potential source of literacy activities.

3. Data relating to this point are available from a number of student projects in Developmental Psychology courses I have supervised. In total these projects have sampled 13 preschools and over 300 children. They have shown that the book-corner is a moderately frequent free-choice source of activity for girls and boys (chosen around 5 per cent of the observed time; which is similar to observations reported in some United States preschools by Dickinson *et al.* 1992). Children can engage in shared book reading and independent book reading during this time. Substantial differences between preschools do occur, however, both in terms of gender differences and in terms of overall usage. For example, in one kindergarten, girls chose 17 per cent of the time and in another boys were not present once over five observation sessions each lasting 45 minutes. The significance of teacher variables in the use of this space has been found by research into setting events for both personal use and tutorial effectiveness of book spaces in preschool and early-grade classrooms. A number of researchers have analysed, for example, the significance of the size of groups and the interactive style of the teacher (Dickinson *et al.* 1992; Morrow 1988).

4. The nature of contact with school in the early grades and its relationships with school achievement have been explored in a number of studies. The relationships are complex and depend on the nature and reasons for contact as well as who initiates them. But in general, the frequency of contacts, particularly continuing and informative contact, is related to achievement. What is clear is that what transpires in the contact in terms of information affecting beliefs and strategies is significant (see McNaughton 1987; Snow *et al.* 1991; Tizard *et al.* 1988).

5. These summaries of school-book reading at home come from descriptions of standard practices at home and at school as well as experimental programmes which have modified the procedures for hearing children read at home (Glynn *et al.* 1989; Glynn & McNaughton 1985; McNaughton 1987; McNaughton & Ka'ai 1990; McNaughton *et al.* 1992b).

In addition to the research in New Zealand there is a substantial body of international research information from interventions which have focused on parents hearing their children read at home (e.g. Hewison 1988). In each case the experimental evidence supports the findings that hearing reading at home can have significant effects on learning to read at school, given the presence of certain boundary conditions. These include the degree of similarity between activities at home and school, the clarity of the information and feedback for parents, and the quality of the ongoing collaboration between parents and teachers.

6. Typically, when a child reads at school the tutorial configuration is a complex mixture of collaborative participation and item conveyancing. Teachers anticipate and respond to meanings in the accurate oral reading. They use embedded correction routines which support the reader's strategic focus on meaning and syntax while highlighting critical information about text components (McNaughton 1987).

7. In their study of home and school influences on literacy, Snow and her colleagues introduced the concept of resilience (Snow *et al.* 1991). They examined the influence of a number of variables, including stress at home, on achievement levels at school. The financial and social stress experienced by parents in low-income families was not directly related to achievement in reading and writing at school. Differences between families were related more to the psychological and social resources the families had for coping with stress (such as strategies for managing intra-family conflict) than to the amount of stress as such. The researchers also note, however, that direct relationships would have been affected because a number of families did not participate in the study, or dropped out because of family problems.(see also Fox 1990).

8. The availability of appropriate materials such as books is a critical resource which, in turn, is dependent on other resources available to the family. In the SOL study, library trips were described as a major activity for the Samoan families and the Maori families. Libraries were major resources for ambient, joint, and independent literacy activities within these families. Yet at the time we were carrying out this study some libraries were being closed by the local authorities as cost-cutting exercises. From the families' point of view this would have been a devastating occurrence.

Glossary

Activities Everyday events for which the participants have ideas and goals that define purposes and intentions and guide actions. What participants do within activities is patterned. That is, there are typical or habitual ways of carrying out the activity. When parents read storybooks with 4-year-olds, patterns can be found in the interactions that take place. Participation also is strategic to meet the goals of the activity and everyday demands (see note 3, Chapter Four).

Ambient activities are those activities in which a person does not engage directly as a participant but which can be observed happening in their everyday lives.
Joint activities are those in which the learner engages directly with another or others.
Personal activities are those in which the learner engages in relatively independently so that the ideas and goals and performances are under the learner's control.

Co-constructivism (-ionist; -ionism) A theory of psychological development which explains development as a product of dynamic, mutual and interdependent constructions of an active learner and social and cultural processes. For example, the participation patterns and the focus of interactions when books are read to children in families carry meaning about social and cultural identity. Together, these forces activate learning, and define and channel development. The theory is a contemporary elaboration of sociocultural or sociohistorical explanations of psychological development such as those outlined by Lev Vygotsky.

Configurations of tutorials These are variations in the way that properties of teaching (including self-teaching) and learning are expressed within tutorial systems that develop within activities. For example, properties such as the way in which adjustments are made by the tutor in the degree of support or guidance, the ways in which performances are highlighted, and the explicitness of the support provided for the learner can be expressed in different ways. Three general configurations are described in Chapter 3. These tutorials are called *collaborative participation*, *directed performance*, and *item conveyancing*.

Dyadic, triadic, multiparty These terms describe the number of participants in an activity. Dyadic refers to two participants (e.g. mother and child); triadic refers to three participants (e.g. mother, older child, and younger child); multiparty refers to three or more participants (e.g. mother, auntie, older child, and younger child).

Ecology The ecology of children's development comprises the functional properties of the environment within which children develop. The focus on ecology draws attention to how properties of environments channel or provide opportunities for developmental processes. A general model has been provided by Bronfenbrenner (see Chapter 9).

Expertise An expert is someone skilled in an activity. Becoming an expert entails the development and integration of skilful behaviour, including strategies for performing, self-regulation, and a knowledge base that includes knowing the purposes, values, and roles associated with the activity.

Situated expertise refers to the nature of expertise developing within a particular activity. Expertise reflects the form the activity has taken and, to a greater or lesser extent, is specific to that activity and those which have some similarity. Children who become expert in the activity of reading for narrative meanings (among other things) have developed knowledge about narrative properties of storybooks, and strategies for comprehending and checking these (see Chapter 5, p. 92).

Pedagogy This term refers not only to processes of teaching and learning, but also to the underlying assumptions about the nature of teaching and learning. A pedagogy of enquiry, for example, might describe (among other things) the processes through which the learner's enquiring and problem solving is facilitated, and also the assumptions about the nature of children as constructive learners.

Practices These involve '… activities for which the culture has normative expectations of the form, manner, and order of conducting repeated or customary actions requiring specified skills and knowledge … Cultural practices have to be learned as systems of activities.' (Laboratory of Comparative Human Cognition, 1983, p. 333). Practices are the larger groupings of activities, such as reading for narrative meaning and writing narratives, which define the general nature of, for example, literacy within the family.

Scaffold (scaffolded learning/teaching/instruction) This metaphor originally was introduced by Wood, Bruner, & Ross (1976) to describe the functions of tutoring in a problem-solving task. It has been elaborated on by other researchers as a model for describing co-construction processes at work in family language and literacy activities. In this book it has been employed to describe tutorial properties in systems for learning and development (Chapter 4). These distinctive properties (see Chapter 3) come from the original metaphor which characterised support as contingent, dynamic, and progressive.

Socialisation (in the family) Processes whereby young members in the family are reared to become expert members of the social and cultural groups to which the family belongs.

Strategy A strategy is defined by Wood (1978) as a 'programmatic assembly of operations aimed at a common goal'. The important features of strategic performance are flexible sequences of acts which are directed at attaining a particular goal. Acting strategically entails the potential availability and selection of alternative routes to the goal. When trying to understand events in a storybook, a child who is listening may act in a number of ways to clarify meaning, for example, by asking the reader, or by using what has been read, predicting what will happen as the story is read, and integrating information from those sources.

System Generally defined, a system is a complex whole made up of a set of interconnected parts. The properties of the whole are greater than the individual parts. A system is goal oriented and dynamic, responding to internal and external pressures and stimuli to maintain (or even modify) goals. The concept can be applied to the family; for example, starting early-childhood education or school creates new pressures and triggers for the family. These include how to respond to new information about child rearing and education in ways that maintain the family's own goals for the child. The concept of system can be applied at a more focused level, too, to the properties of participation and interaction within an activity which enable learning and development to take place (see Chapter 4, p. 62). It can also be applied at more encompassing ecological levels as in Bronfenbrenner's model of the ecology of human development (see Chapter 9, page 162).

Tutorial systems These are systems of learning and development that evolve in joint activities, forms of which take place in personal activities, too (see Chapter 4, p. 62). They are vehicles for teaching and learning in activities and have three general features (described in Chapter 4). They are based on activities, they can be organised in a variety of ways to enable expertise to be co-constructed (tutorial configurations), and through the interaction relationships between participants are formed and developed. Systems of learning and development form in the activity of writing a name. Family members interact with the child during the activity in ways that provide a tutorial structure (e.g. a performance-directed tutorial). The child already has relationships with these family members and these relationships are expressed further and developed in the system.

Zones of development As applied to family socialisation this concept refers to the sets of experiences and opportunities which family environments contain. These shape development by determining its immediate direction. They carry the family's socialisation processes. Valsiner (1987) proposed two such general Zones, which were additions to the one described by Vygotsky as the Zone of Proximal Development.

Zone of Proximal Development   This is defined by the difference between what a learner can do independently or can do with guidance and collaboration with more expert persons. See Chapter 4, p. 64. Valsiner's additions were:

Zones of Freedom of Movement   These contain the set of possibilities for activity and thus frameworks for action (and expertise) in particular settings; for example, a bedroom may contain books.

Zones of Promoted Actions   These are the particular objects, areas, ways of acting, and activities which are promoted or highlighted within a Zone of Freedom of Movement. Family members may highlight the books within the Zone of Freedom of Movement by reading them to the child and so create a Zone of Promoted Action.

References

Anderson, R. C. Wilson, P. T., & Fielding, L. G. (1988). 'Growth in reading and how children spend their time outside of school'. *Reading Research Quarterly*, XXIII, 3, 285–303.

Ansley, M. M. (1990). 'The way we are'. *Listener*, Sept. 3, 20–22.

Aries, P. (1962). *Centuries of Childhood: A Social History of Family Life.* Vintage Books, New York.

Bakhtin, M. (1981). *The Dialogic Imagination.* The University of Texas Press, Austin.

Barney, D. (1975). *Who Gets to Preschool?* New Zealand Council for Educational Research, Wellington.

Bishop, R. & Glynn, T. (1992). 'He kanohi kitea: conducting and evaluating educational research'. *New Zealand Journal of Educational Studies*, 27, 2, 125–136.

Bissex, G. (1980). *Gnys at Wrk: A Child Learns to Write and Read.* Harvard University Press, Cambridge, MA.

Blatchford, P. (1991). 'Children's writing at 7 years: associations with handwriting on school entry and pre-school factors'. *British Journal of Educational Psychology* vol. 61, 73-84.

Bohr, N. H. D. (1972). *Collected Works.* North Holland Publishing Co., Amsterdam.

Bornstein, M. H. (1991). 'Approaches to parenting in culture'. In M. H. Bornstein (ed.). *Cultural Approaches to Parenting.* Lawrence Erlbaum, Hillsdale, NJ.

Bourdieu, P. (1973) 'Cultural reproduction and social reproduction'. In R. Brown (ed.). *Knowledge, Education and Social Change.* Tavistock, London.

Bransford, J. D. (1979). *Human Cognition: Learning, Understanding and Remembering.* Wadsworth, Belmont, CA.

Bronfenbrenner, U. (1979). *The Ecology of Human Development.* Harvard University Press, Cambridge, MA.

Bronfenbrenner, U. (1986). 'Ecology of the family as a context for human development'. *Developmental Psychology*, 22, 723–742.

Britton, B. K. & Pellegrini, A. D. (1990). *Narrative Thought and Narrative Language.* Lawrence Erlbaum, Hillsdale, NJ.

Brown, A. L. & Palincsar, A. S. (1990). 'Guided, co-operative learning and individual knowledge acquisition'. In L. Resnick (ed.). *Knowledge, Learning and Instruction: Essays in Honor of Robert Glaser.* Lawrence Erlbaum, Hillsdale, NJ.

Bruner, J. S (1983). *Child's Talk.* Norton, New York.

Canetti, E. (1977). *The Tongue Set Free.* Picador, London.

Carey, S. & Gelman, R. (1991) (eds). *The Epigensis of Mind: Essays on Biology and Cognition.* Lawrence Erlbaum, Hillsdale, NJ.

Carroll, L. (1986). *The Complete Illustrated Works of Lewis Carroll.* Chancellor Press, London.

Cazden, C. (1988). *Classroom Discourse.* Heinemann, Portsmouth, NH.

Cazden, C. (1993a). 'Vygotsky, Hymes, & Bakhtin: from word to utterance and voice'. In Forman, E. A., Minick, N., & Stone, C. A. (eds.) *Contexts from Learning: Sociocultural Dynamics in Children's Development.* Oxford University Press, New York.

Cazden, C. (1993b) 'Immersing, revealing and telling: a continuum from implicit to explicit teaching'. Paper presented to the Second International Conference on Teacher Education in Second Language Teaching. City Polytechnic of Hong Kong.

Clarke-Stewart, K. A. (1978). 'Popular primers for parents'. *American Psychologist*, 33, 359–69.

Clay, M. M. (1966). 'Emergent reading behaviour'. Unpublished Ph.D. thesis, University of Auckland.

Clay, M. M. (1975). *What Did I Write?* Heinemann, Auckland.

Clay, M.M. (1979a) *Reading: The Patterning of Complex Behaviour.* (2nd edn.), Heinemann, Auckland.

Clay, M. M. (1979b). *The Early Detection of Reading Difficulties: A Diagnostic Survey with Recovery Procedures.* Heinemann, Auckland.

Clay, M. M. (1985). 'Engaging with the school system'. *New Zealand Journal of Educational Studies,* 20, 20–38.

Clay, M. M. & Cazden, C. B. (1990). 'A Vygotskian perspective on reading recovery'. In L. Moll (ed.). *Vygotsky and Education.* Cambridge University Press, Cambridge.

Cole, M. & Cole, S. R. (1989). *The Development of Children.* Scientific American Books, New York.

Cram, F. (1993) (ed.). 'Issues — Maori research'. *Bulletin of the New Zealand Psychological Society,* 76.

Cummins, J. (1986). 'Empowering minority students: a framework for intervention'. *Harvard Educational Review.* 56, 18–36.

Czerniewska, P. (1992). *Learning about Writing.* Blackwell, Oxford.

Damon, W. (1991). 'Problems of direction in socially shared cognition'. In L. B. Resnick, J. M. Levine & S. D. Teasdale (eds.). *Perspectives in Socially Shared Cognition.* American Psychological Association, Washington DC.

Dickinson, D. K., De Temple, J. M., Hirschler, J. A., & Smith, M. W. (1992). 'Book Reading with Preschoolers: Co-construction of text at home and at school'. *Early Childhood Research Quarterly,* 7, 323–346.

Donaldson, M. (1978). *Children's Minds.* Fontana, Glasgow.

Duranti, A. & Ochs, E. (1986). 'Literacy instruction in a Samoan village'. In B. B. Schieffelin & P. Gilmore (eds.). *The Acquisition of Literacy: Ethnographic Perspectives.* Ablex, Norwood, NJ.

Durkin, D. (1966). *Children who Read Early.* Teachers College Press, New York.

Dyson, A. H. (1991). 'Cultural bridges, literacy towers and kids carnivals: a child's perspective of cultural traditions and school success'. Paper presented at the the 16th National conference of the Australian Reading Association, Adelaide, July.

Dyson, A. H. & Freedman, S. (1991). 'Writing'. In J. Flood, D. Lapp, & J. R. Squire (eds.). *Handbook of Research on Teaching the English Language Arts.* Macmillan, New York.

Elley, W. B. (1985). 'What do children learn from being read to?' *Set 1,* Item 11. New Zealand Council for Educational Research.

Elley, W. B. (1992). *How in the World do Students Read?* The International Association for the Evaluation of Educational Achievement, New York.

Erickson, F. (1984). 'School literacy, reasoning and civility: an anthropologist's perspective'. *Review of Educational Research,* 54, 525–546.

Fairburn-Dunlop, P. (1984). 'Factors associated with language maintenance: the Samoans in New Zealand'. *New Zealand Journal of Educational Studies*, 19, 2, 99–113.

Feldman, C. F. (1991). 'Oral metalanguage'. In D. R. Olson & N. Torrance (eds.). *Literacy and Orality.* Cambridge University Press, Cambridge.

Ferreiro, E. (1985). 'Literacy development: a psychogenetic perspective'. In D. R. Olson, N. Torrance, & A. Hildyard (eds.). *Literacy, Language and Learning: The Nature and Consequences of Reading and Writing.* Cambridge University Press, Cambridge.

Ferreiro, E. & Teberosky, A. (1982). *Literacy before Schooling.* Heinemann, Exeter, NH.

Foote, L. & Regget, L. (1992).'Perspectives on literacy'. Unpublished paper, University of Otago.

Fox, B. J. (1990). 'Antecedents of illiteracy'. *Social Policy Report*, IV, No. 4.

Gallimore, R. & Goldenberg, C. (1993). 'Activity settings of early literacy: home and school factors in children's emergent literacy'. In Forman, E. A., Minick, N. & Stone, C. A. (eds.). *Contexts from Learning: Sociocultural Dynamics in Children's Development*. Oxford University Press, New York.

Gardner, H. (1991). *To Open Minds*. Basic Books.

Gee, J. P. (1990). *Social Linguistics and Literacies: Ideology in Discourses*. The Falmer Press, London.

Glynn, T., Crook, T., Bethune, N., Ballard, K., & Smith, J. (1989). 'Reading recovery in context'. Final Report to the New Zealand Department of Education. University of Otago, Dunedin.

Glynn, T. & McNaughton, S. (1985).'The Mangere Home and School Remedial Reading Procedures: continuing research on their effectiveness'. *New Zealand Journal of Psychology*, 14, 66–77.

Goodman, Y. (1990). *How Children Construct Literacy*. International Reading Association, Newark.

Goodnow, J. J. & Collins, W. A. (1990). *Development According to Parents: The Nature, Sources, and Consequences of Parents' Ideas*. Lawrence Erlbaum, Hillsdale, NJ.

Goodridge, M. J & McNaughton, S. (1993). 'The co-construction of writing expertise in family activity'. Paper presented at the 15th Annual conference of the New Zealand Association for Research in Education, Hamilton, 2-5 December.

Goodridge, M. J. & McNaughton, S. (1994). 'How families and children construct writing expertise before school: an analysis of activities in Maori, Pakeha and Samoan families'. Paper presented at the 20th Annual conference of the New Zealand Reading Association, 8-11 May, Auckland

Goswami, U. & Bryant, P. (1990). *Phonological Skills and Learning to Read*. Lawrence Erlbaum, London.

Gould, S. J. (1989). *Wonderful Life*. Penguin Books, London.

Graham, S. (1992). '"Most of the subjects were white and middle class": trends in published research on African Americans in selected APA Journals, 1970-1989'. *American Psychologist*, 47, 5, 629-639.

Gray, J. D. (1994). 'Is there evidence to show that peer groups help to construct writing expertise?' Unpublished M.A. thesis, University of Auckland.

Greenfield, P. M. (1984). 'A theory of the teacher in the learning activities of everyday life'. In B. Rogoff & J. Lave (eds.). *Everyday Cognition: Its Development in Social Context*. Harvard University, Cambridge, MA.

Guthrie, J. T. (1982). 'Reading in New Zealand: achievement and volume'. *Reading Research Quarterly*, 17, 6-27.

Guthrie, J. T. & Greaney, V. (1991). 'Literacy acts'. In P. D. Pearson, R. Barr, M. L. Kamil, & P. Mosenthal (eds.). *Handbook of Reading Research. Vol 2*. Longman, New York.

Heath, S. B. (1982). 'What no bedtime story means: narrative skills at home and at school'. *Language in Society*, 11, 49-76.

Heath, S. B. (1983). *Ways with Words: Language, Life and Work in Communities and Classrooms*. Cambridge University Press, Cambridge.

Heath, S. B. & Branscombe, A. (1986). 'The book as a narrative prop in language acquisition'. In B. Schieffelin & P. Gilmore (eds.). *The Acquisition of Literacy: Ethnographic Perspectives*. Ablex, Norwood, NJ.

Hendricks, A., Meade, A., & Wiley, C. (1993). *Competent Children: Influences of Early Childhood Experiences*. New Zealand Council for Educational Research, Wellington.

Hewison, J. (1988). 'The long term effectiveness of parental involvement in reading: a follow-up to the Haringey Reading Project'. *British Journal of Educational Psychology*, 58, 184–190.

Hildreth, G. (1936). 'Developmental sequences in name writing'. *Child Development*, 7, 291–303.

Hohepa, M., Smith, G. H., Smith, L. T., & McNaughton, S. (1992). 'Te Kohanga Reo hei tikanga ako i te Reo Maori: Te Kohanga Reo as a context for language learning'. *Educational Psychology*, 12, 3 & 4, 323–346.

Irwin, K. (1992). 'Maori research methods and processes: an exploration and discussion'. Paper presented at the 2nd Joint New Zealand and Australian Associations for Research in Education conference, Geelong, November.

Johnston, R. (1983). *A Revision of Socioeconomic Indices for New Zealand*. New Zealand Council for Educational Research, Wellington.

Jones, A. (1991). *'At School I've Got a Chance.' Culture/Privilege: Pacific Islands and Pakeha Girls at School*. Dunmore Press, Palmerston North.

Kempton, M. (1994). 'Learning to write one's name: developmental processes in an emergent writing activity'. Unpublished M.A. thesis, University of Auckland.

Kessen, W. (1979). 'The American child and other cultural inventions'. *American Psychologist*. 34, 815–820.

Kessen, W. (1991). 'Commentary: dynamics of enculturation'. In M. H. Bornstein (ed.). *Cultural Approaches to Parenting*. Lawrence Erlbaum, Hillsdale, NJ.

Laboratory of Comparative Human Cognition (1983). 'Culture and cognitive development'. In P. H. Mussen (ed.). *Handbook of Child Psychology (Vol. 1: History, Theory and Methods)*. Wiley, New York.

Laosa, L. M. (1989). 'Social competence in childhood: toward a developmental, socioculturally relativistic paradigm'. *Journal of Applied Developmental Psychology*, 10, 447–461

Lave, J. & Wenger, E. (1991). *Situated Learning: Legitimate Peripheral Participation*. Cambridge University Press, Cambridge.

Learning Media (1992). *Dancing with the Pen: The Learner as a Writer*. Ministry of Education, Wellington.

Leont'ev, A. N. (1981). 'The problem of activity in psychology'. In J. V. Wertsch, (ed.). *The Concept of Activity in Soviet Psychology*. Sharpe, Armonk, New York.

Lindfors, J. W. (1987). *Children's Language and Learning* (2nd edn.). Prentice-Hall, Englewood Cliffs, NJ.

McDonald, G. (1970). 'Preschool education'. In R. J. Bates (ed.). *Prospects in New Zealand Education*. Hodder & Stoughton, Auckland.

McMillan, B. (1983). 'Parents' beliefs about some aspects of children's development'. Paper presented at the 5th National conference of the New Zealand Association for Research in Education, Wellington, December.

McNaughton, S. (1987). *Being Skilled: The Socializations of Learning to Read*. Methuen, London.

McNaughton, S. (1991). 'The faces of instruction: models of how children learn from tutors'. In J. Morss & T. Linzey (eds.). *Growing Up: The Politics of Human Learning*. Longman, Auckland.

McNaughton, S. (1992). 'Lines of communication: the psychology of homework'. Paper presented at the New Zealand Reading Association Annual Conference, Wellington, May.

McNaughton, S. (1994a) 'Human development and the reconstruction of culture: a commentary on Valsiner (1994)'. In P. Van Geert & L. Mos (eds.). *Annals of Theoretical Psychology,* vol.x. Plenum, New York.

McNaughton, S. (1994b). 'Why there might be several ways to read storybooks to pre-schoolers in Aotearoa/New Zealand: models of tutoring and sociocultural diversity in how families read books to preschoolers'. In M. Kohl de Oliviera & J. Valsiner (eds.). *Literacy in Human Development.* Ablex, Norwood, NJ.

McNaughton, S., Glynn, T., & Robinson, V. (1987). *Pause, Prompt and Praise: Effective Tutoring for Remedial Education.* Positive Products, Birmingham.

McNaughton, S. & Ka'ai, T. (1990). 'Two studies of transitions: socializations of literacy and Te hiringa take take: Mai i Te Kohanga Reo ki te kura'. Report to the New Zealand Ministry of Education. Education Department, University of Auckland.

McNaughton, S., Kempton, M., & Turoa, L. (1994). 'Tutorial configurations in an early writing activity at home'. Poster presented at the XIIIth Biennial meeting of the International Society for the Study of Behavioural Development. Amsterdam, 28 June–2 July.

McNaughton, S., Ka'ai, T., & Wolfgramm, E. (1993). 'The perils of scaffolding: models of tutoring and sociocultural diversity in how families read storybooks to preschoolers'. Paper presented at the biennial conference of the Society for Research in Child Development, New Orleans, March.

McNaughton, S., Parr, J., Timperley, H., & Robinson, V. M. J. (1992a). 'A report to the Ministry of Education: SA survey of community and school educational values'. Auckland Uniservices Ltd, Auckland.

McNaughton, S., Parr, J., Timperley, H. & Robinson, V. M. J. (1992b). 'Beginning reading and sending books home to read: a case for some fine tuning'. *Educational Psychology,* 12, 3 & 4, 239–247.

Mason, J. M. & Allen, J. (1986). 'A review of emergent literacy with implications for research and practice in reading'. In E. Z. Rothkopf (ed.). *Review of Research in Education,* vol. 13. American Educational Research Association, Washington.

Metge, J. (1984). *Learning and Teaching: He Tikanga Maori.* New Zealand Ministry of Education, Wellington.

Meyerhoff, M. K. & White, B. L. (1986). 'New parents as teachers'. *Educational Leadership,* 42–46.

Ministry of Education (1993). *Maori in Education.* Ministry of Education, Wellington.

Morrow, I. M. (1988). 'Young children's responses to one-to-one story readings in school settings'. *Reading Research Quarterly,* 23, 89–107.

Morss, J. (1991). 'After Piaget: rethinking "cognitive development"'. In J. Morss & T. Linzey (eds.). *Growing Up: The Politics of Human Learning.* Longman, Auckland.

Murray, S. (1974). *Te Karanga a te Kotuku.* Maori Organisation on Human Rights, Wellington.

Nalder, S. (1985). 'Emergent reading'. Unpublished report. Reading Advisory Service, Auckland.

Nash, R. (1991). 'In defense of a common curriculum and a universal pedagogy'. In J. R. Morss & T. Linzey (eds.). *Growing Up: The Politics of Human Learning.* Longman Paul, Auckland.

Nash, R. (1992). 'Review essay'. *New Zealand Sociology,* 7, 2, 73–97.

Nash, R. & Harker, R. (1990). 'Working with class: the educational expectations and practices of class-resourced families'. Paper presented at the annual conference of the New Zealand Association for Research in Education, Auckland, December.

National Institute of Education, (1985). *Becoming a Nation of Readers.* The Report of the Commission on Reading. US Department of Education, Washington, DC.

Nepe, T. M. (1990). 'Te toi huarewa tipuna'. Unpublished M.A. thesis, University of Auckland.

Newman, D., Griffin, P., & Cole, M. (1989). *The Construction Zone: Working for Cognitive Change in School.* Cambridge University Press, Cambridge.

Nicholson, T. (1979). 'What parents know about reading — and what we need to tell them'. Paper presented at the Australia and New Zealand Association for the Advancement of Science Congress, Auckland, January.

Ninio, A. (1980). 'Picture-book reading in mother-infant dyads belonging to two subgroups in Israel'. *Child Development,* 51, 587–590.

Ninio, A. & Bruner, J. S. (1978). 'The achievement and antecedents of labelling'. *Journal of Child Language,* 5, 5–15.

Ochs, E. (1982). 'Talking to children in Western Samoa'. *Language in Society,* 11, 77–104.

Ogbu, J. U. (1991). 'Cultural mode, identity and literacy'. In J. W. Stigler, R. A. Shweder, & G. Herdt (eds.). *Cultural Psychology: Essays on Comparative Human Development.* Cambridge University Press, Cambridge.

Olson, D. R. (1991). 'Literacy as a metalinguistic activity'. In D. R. Olson & N. Torrance (eds.). *Literacy and Orality.* Cambridge University Press, Cambridge.

Olson, D. R. & Torrance, N. (1991) (eds.). *Literacy and Orality.* Cambridge University Press, Cambridge.

Pankhurst, F. (1989). 'The acquisition of cartography in preschool children'. Unpublished Ph.D. thesis, Victoria University of Wellington.

Pappas, C. C. (1993). 'Is narrative "primary"? Some insights from kindergarteners' pretend readings of stories and information books'. *Journal of Reading Behavior,* 25, 97–129.

Pellegrini, A. D., Perlmutter, J. C., Galda, L., & Brody, G. H. (1990). 'Joint book reading between black head start children and their mothers'. *Child Development,* 61, 443–453.

Pere, R. R. (1991). *Te Wheke.* Ao Ako Global Learning New Zealand Limited, Gisborne.

Peters, A. M. & Boag, S. T. (1987). 'Interactional routines as cultural influences upon language acquisition'. In B. B. Schieffelin & E. Ochs (eds.). *Language Socialization across Cultures.* Cambridge University Press, Cambridge.

Phillips, G. (1986). 'Storybook reading to children in their home environment'. Unpublished M.A. thesis, University of Auckland.

Phillips, G. & McNaughton, S. (1990). 'The practice of storybook reading to preschool children in mainstream New Zealand families'. *Reading Research Quarterly,* 25, 3, 196–212.

Piaget, J. (1970). 'Piaget's theory'. In P. H. Mussen (ed.). *Carmichael's Handbook of Child Psychology.* (3rd edn., vol. 1). Wiley, New York.

Podmore, V. N. & Bird, L. (1991). *Parenting and Children's Development in New Zealand: Research on Childrearing, Parental Beliefs and Parent Child Interaction.* 'State-of-the-Art', Monograph No. 3. New Zealand Association for Research in Education.

Reed, E. S. (1993). 'The intention to use a specific affordance: a conceptual framework for psychology'. In R. H. Wozniak & K. W. Fischer (eds.). *Development in Context: Acting and Thinking in Specific Environments.* Lawrence Erlbaum, Hillsdale, NJ.

Renshaw, P. D. (1992). 'Reflecting on the experimental context: parent's interpretations of the education motive during teaching episodes'. In L. T. Winegar & J. Valsiner (eds.). *Children's Development within Social Context.* Lawrence Erlbaum, Hillsdale, NJ.

Rogoff, B. (1990). *Apprenticeship in Thinking: Cognitive Development in Social Context.* Oxford University Press, Oxford.

Rogoff, B. (1993). 'Children's guided participation and participatory appropriation in sociocultural sctivity'. In R. H. Wozniak and K. W. Fischer (1993) (eds). *Development in Context: Acting and Thinking in Specific Environments.* Lawrence Erlbaum, Hillsdale, NJ.

Rogoff, B. & Lave, J. (1984) (eds). *Everyday Cognition: Its Development in Social Context.* Harvard University Press, Cambridge, MA.

Rose, G. (1977). *Ahhh Said the Stork.* Faber, London.

Sameroff, A. J. (1982). 'Development and the dialectic: the need for a systems approach'. In W. A. Collins (ed.). *The Concept of Development. The Minnesota Symposium of Child Psychology.* Lawrence Erlbaum, Hillsdale, NJ.

Schickedanz, J. (1984). 'A study of literacy events in the homes of six preschoolers'. Paper presented at the 34th National Reading conference. Florida, November 30.

Schieffelin, B. B. & Ochs, E. (1986) (eds.). *Language Socialization Across Cultures.* Cambridge University Press, Cambridge.

Scribner, S. and Cole, M. (1981). *The Psychology of Literacy.* Harvard University Press, Cambridge, MA.

Sinclair, K. (1991). *Kinds of Peace: Maori People after the Wars 1870–85.* Auckland University Press, Auckland.

Slobin, D. I. (1990). 'The development from child speaker to native speaker'. In J. W. Stigler, R. A. Shweder, & G. Herdt (eds.). *Cultural Psychology: Essays in Comparative Human Development.* Cambridge University Press, Cambridge.

Smith, F. (1978). *Understanding Reading.* (2nd edn.). Holt, Rinehart & Winston, New York.

Smith, G. (1987). 'Akonga Maori: preferred Maori teaching and learning methodologies'. Research Unit for Maori Education, University of Auckland, Auckland.

Smith, L. (1991). 'Te Rapunga i te Ao Marama: Maori perspectives on Research in Education'. In J. R. Morss & T. J. Linzey (eds.). *The Politics of Human Learning: Human Development and Educational Research.* Longman Paul, Auckland.

Smith, L.T., Smith, G. H., & McNaughton, S. (1989). 'Some implications for research within Kura Kaupapa Maori classrooms'. Paper presented at the annual conference of the New Zealand Association for Research in Education.

Snow, C. E. (1983). 'Literacy and language: relationships during the preschool years'. *Harvard Educational Review*, 55, 165–189.

Snow, C. E. (1986). Interview. *Baby Talk.* Open University, Video tape, E362–O6A.

Snow, C. E. (1991) 'The theoretical basis for relationships between language and literacy in development'. *Journal of Research in Childhood Education*, 6, 1, 5–10.

Snow, C. E., Barnes, W. S., Chandler, J., Goodman, I. F., & Hemphill, L. (1991). *Unfulfilled Expectations: Home and School Influences on Literacy.* Harvard University Press, Cambridge, MA.

Stanovich, K. F. (1992). 'Differences in reading acquisition: causes and consequences'. *Reading Forum NZ,* 3, 3–21.

Stein, N. I. & Glenn, C. G. (1979). 'An analysis of story comprehension in elementary school children'. In R. Freedle (ed.). *New Directions in Discourse Processing (Vol. 2).* Ablex, Norwood, NJ.

Stevenson, H. W. & Lee, S. (1990). 'Contexts of achievement'. *Monographs of the Society for Research in Child Development.* 55, nos. 1–2.

Sulzby, E, & Teale, W. (1991). 'Emergent literacy'. In P. D. Pearson, R. Barr, M. L. Kamil, & P. Mosenthal (eds.). *Handbook of Reading Research. Vol 2.* Longman, New York.

Tagoilelagi, F. (1992). 'Observations of Lotu'. Unpublished manuscript, University of Auckland, Auckland.

Tangaere, A. R. & McNaughton, S. (1994). 'From preschool to home: processes of generalisation in language acquisition from an indigenous language recovery programme'. *International Journal of Early Years Education,* 2 (1), 23–40.

Teale, W. H. (1984). 'Reading to young children: its significance for literacy development'. In H. Goelman, A. Oberg, & F. Smith (eds.). *Awakening to Literacy.* Heinemann, Exeter, NH.

Thackery, S., Syme, K., & Hendry, D. (1992). *A Survey of School Entry Practices: How Schools Gather Information on New Entrants.* Learning Media, Ministry of Education, Wellington.

Tharp, R.G. & Gallimore, R. (1988). *Rousing Minds to Life: Teaching, Learning and Schooling in Social Context.* Cambridge University Press, Cambridge.

Tizard, B., Blatchford, P., Burke, J., Farquhar, C., & Plewis, I. (1988). *Young Children at School in the Inner City.* Lawrence Erlbaum, London.

Troughton, J. (1977). *What Made Tiddalik Laugh?* Nelson, Melbourne.

Valsiner, J. (1987). *Culture and the Development of Children's Action.* Wiley, Chichester.

Valsiner, J. (1988). 'Ontogeny of co-construction of culture within socially organized environmental settings'. In J. Valsiner (ed.). *Child Development within Culturally Structured Environments.* Vol. 2. Ablex, New Jersey.

Valsiner, J. (1994a). 'Culture and human development: a co-constructivist perspective'. In P. Van Geert & L. Mos (eds.). *Annals of Theoretical Psychology.* vol. x. Plenum, New York.

Valsiner, J. (1994b). 'Co-constructivism: what is (and is not) in a name?' In P. Van Geert & L. Mos (eds.). *Annals of Theoretical Psychology,* vol. x. Plenum, New York.

Valsiner, J. & Van der Veer, R. (1993). 'The encoding of distance: the concept of the "Zone of Proximal Development" and its interpretations'. In R. R. Cocking & K. A. Renninger (eds.). *The Development and Meaning of Psychological Distance.* Lawrence Erlbaum, Hillsdale, NJ.

Van der Veer, R. & Valsiner, J. (in press) (eds.). *The Vygotsky Reader.* Basil Blackwell, Oxford.

Vasconcellos, V. M. R. & Valsiner, J. (1993). 'From imitation to cognitive construction: contrasting Wallon and Piaget'. Paper presented at the Fifth Conference of the International Society for Theoretical Psychology, Chateau de Bierville, France, April.

Vygotsky, L. S. (1978). *Mind in Society: The Development of Higher Psychological Processes.* (M. Cole, V. John-Steiner, S. Scribner, & E. Souberman, eds. & trans.). Harvard University Press, Cambridge, MA.

Wade, W. R. (1838). *Journey in the North Island of New Zealand.* Reprint. Capper Press, Christchurch.

Wagemaker, H. (1991). IEA Reading Literacy Study: Preliminary Analysis — New Zealand. Bulletin: Research and Statistics Division New Zealand Ministry of Education, 63–73.

Wagner, D. A. & Spratt, J. E. (1987). 'Cognitive consequences of contrasting pedagogies: the effects of Quranic pre-schooling in Morocco'. *Child Development,* 58, 1207–1219.

Walkerdine, V. & Lucy, H. (1989). *Democracy in the Kitchen.* Virago Press, London.

Wells, G. (1985). 'Preschool literacy-related activities and success in school'. In D. Olson, M. Torrance, & A. Hildyard (eds.). *Literacy, Language and Learning.* Cambridge University Press, London.

Wells, G. (1986). 'The language experience of five-year-old children at home and at school'. In J. Cook-Gumperz (ed.). *The Social Construction of Literacy.* Cambridge University Press, Cambridge.

Wendt, A. (1986) 'Exam failure praying'. In *The Birth and Death of the Miracle Man: A Collection of Short Stories.* Viking, New York.

Wertsch, J. V. (1985). (ed.). *Vygotsky and the Social Formation of Mind.* Harvard University Press, Cambridge, MA.

Wertsch, J. V. (1991). *Voices of the Mind: A Sociocultural Approach to Mediated Action.* Harvard University Press, Cambridge, MA.

Whitehurst, G. J., Falco, F. L., Lonigan, C. J., Fischel, J. E., DeBaryshe, B. D., Valdez-Menacha, M. C., & Caufield, M. (1988). 'Accelerating language development through picture book reading'. *Developmental Psychology,* 24, 4, 552–559.

Wolf, S. A. & Heath, S. B. (1992). T*he Braid of Literature: Children's Worlds of Reading.* Harvard University Press, Cambridge, MA.

Wolfgramm, E. (1991). 'Becoming literate: the activity of book reading to Tongan preschoolers in Auckland'. Unpublished M.A. thesis, University of Auckland.

Wood, D. (1988). *How Children Think and Learn.* Basil Blackwell, London.

Wood, D., Bruner, J., & Ross, G. (1976). 'The role of tutoring in problem solving'. *Journal of Child Psychology and Psychiatry,* 17, 89–100.

Wozniak, R. H. & Fischer, K. W. (1993) (eds.). *Development in Context: Acting and Thinking in Specific Environments.* Lawrence Erlbaum, Hillsdale, NJ.

Yensen, H., Hague, K., & McCreanor, T. (1989) (eds.). *Honouring the Treaty: An Introduction for Pakeha to the Treaty of Waitangi.* Penguin Books, Auckland.

Index of Authors

Allen, J. 56, 117
Anderson, R. C. 159
Ansley, M. M. 35
Aries, P. 52

Bakhtin, M. 35
Ballard, K. 185, 197
Bethune, N. 185, 197
Bird, L. 53, 181
Bishop, R. 122
Bissex, G. 75, 140
Blatchford, P. 98, 152, 181, 197
Boag, S. T. 84
Bohr, N. H. D. 56
Bornstein, M. H. 53, 91
Bourdieu, P. 18
Bransford, J. D. 78
Branscombe, A. 119, 122
Bronfenbrenner, U. 10, 35, 68, 159, 162–5, 180, 193, 200
Britton, B. K. 101
Brody, G. H. 89, 101, 122
Brown, A. L. 159
Bruner, J. S. 46, 62, 65, 67, 71, 84, 85, 86, 88, 91, 200
Bryant, P. 86, 133
Burke, J. 98, 152, 181, 197

Canetti, E. 49
Carey, S. 12
Carroll, L. 103
Caufield, M. 88, 101
Cazden, C. 32, 67, 70, 71, 101, 102, 166, 175
Clarke-Stewart, K. A. 150
Clay, M. M. 7, 44, 48, 50, 97, 101, 148, 151, 155, 159, 171, 194
Cole, M. 15, 52, 143, 158
Cole, S. R. 52, 158
Collins, W. A. 30, 34, 101
Cram, F. 117
Crook, T. 185, 197

Cummins, J. 166
Czerniewska, P. 13, 50, 51, 56, 127, 142

Damon, W. 78, 118
DeBaryshe, B. D. 88, 101
De Temple, J. M. 26, 31, 111, 120, 181, 197
Dickinson, D. K. 26, 31, 111, 120, 181, 197
Donaldson, M. 158
Duranti, A. 115
Durkin, D. 28, 98, 150, 169
Dyson, A. H. 142, 178

Elley, W. B. 18, 35, 89, 159
Erickson, F. 158

Fairburn-Dunlop, P. 21, 22, 25
Falco, F. L. 88, 101
Farquhar, C. 98, 152, 181, 197
Feldman, C. F. 78
Ferreiro, E. 39, 42, 51, 58, 82, 126, 133
Fielding, L. G. 159
Fischel, J. E. 88, 101
Fischer, K. W. 9
Foote, L. 181
Fox, B. J. 18, 104, 120, 198
Freedman, S. 142

Galda, L. 89, 101, 122
Gallimore, R. 5, 18, 32, 119, 122, 166
Gardner, H. 52
Gee, J. P. 15, 23, 109
Gelman, R. 12
Glenn, C. G. 101
Glynn, T. 122, 185, 197
Goldenberg, C. 18, 119, 122
Goodman, Y. 12, 14, 51, 56, 117
Goodnow, J. J. 30, 34, 101
Goodridge, M. J. 125, 126, 135, 136, 137, 138, 143, 153, 154

Goswami, U. 86, 133
Gould, S. J. 57, 196
Graham, S. 117
Gray, J. D. 75, 173
Greaney, V. 23, 24
Greenfield, P. M. 122
Griffin, P. 158
Guthrie, J. T. 23, 24, 35

Harker, R. 24
Heath, S. B. 23, 31, 39, 56, 71, 90, 110, 111, 118, 119, 122, 127, 129, 141, 147, 153, 154, 172
Hendricks, A. 159, 160
Hendry, D. 147, 172, 178
Hewison, J. 166, 197
Hildreth, G. 51, 57
Hirschler, J. A. 26, 31, 111, 120, 181, 197
Hohepa, M. 32, 116, 166

Irwin, K. 54

Johnson, R. 19

Ka'ai, T. 19, 31, 32, 104, 115, 116, 127, 136, 190, 194, 197
Kempton, M. 67, 130, 132, 142
Kessen, W. 52, 123

Laboratory of Comparative Human Cognition 20, 200
Laosa, L. M. 170, 171
Lave, J. 9, 123
Learning Media 172
Lee, S. 28
Leont'ev, A. N. 77
Lindfors, J. W. 143
Lonigan, C. J. 88, 101
Lucy, H. 122

McDonald, G. 23
McMillan, B. 27, 127

McNaughton, S. 7, 19, 23, 26, 31, 32, 33, 36, 44, 49, 56, 57, 63, 69, 78, 91, 98, 101, 102, 104, 105, 113, 115, 116, 123, 125, 126, 127, 135, 136, 137, 143, 166, 175, 178, 185, 186, 187, 188, 189, 190, 194, 197, 198
Mason, J. M. 56, 117
Meade, A. 159, 160
Metge, J. 54, 78, 94, 114–15, 116
Meyerhoff, M. K. 104
Ministry of Education 150, 197
Morrow, I. M. 197
Morss, J. 12
Murray, S. 35

Nalder, S. 148
Nash, R. 24, 31, 53, 181
National Institute of Education 152, 159
Nepe, T. M. 32, 54
Newman, D. 158
Nicholson, T. 150
Ninio, A. 62, 65, 67, 71, 85, 86, 88, 89, 101

Ochs, E. 32, 33, 87, 94, 115, 122
Ogbu, J. U. 166, 170, 171
Olson, D. R. 138, 142, 143

Palincsar, A. S. 159
Pankhurst, F. 53
Pappas, C. C. 88, 142
Pellegrini, A. D. 89, 101, 122
Pere, R. R. 54
Perlmutter, J. C. 89, 101, 122
Peters, A. M. 84

Phillips, G. 26, 31, 69, 90, 91, 104, 159, 160
Piaget, J. 12–14
Plewis, I. 98, 152, 181, 197
Podmore, V. N. 53, 181

Reed, E. S. 50, 77
Regget, L. 181
Renshaw, P. D. 101, 120
Robinson, V. 185
Rogoff, B. 9, 14, 15, 45, 56, 62, 77, 78, 123, 161
Rose, G. 90
Ross, G. 65, 84, 91, 200

Sameroff, A. J. 4
Schickedanz, J. 134–5, 139–40
Schieffelin, B. B. 32, 94
Scribner, S. 15, 143
Sinclair, K. 35
Slobin, D. I. 50
Smith, F. 28
Smith, G. 32, 116, 166
Smith, J. 185, 197
Smith, L. T. 116, 166, 175
Smith, M. W. 26, 31, 111, 120, 181, 197
Snow, C. E. 7, 18, 19, 35, 50, 62, 111, 142, 181, 197, 198
Spratt, J. E. 94
Stanovich, K. F. 101, 157
Stein, N. I. 101
Stevenson, H. W. 28
Sulzby, E. 7, 26, 39, 65, 88, 92, 104, 105, 108, 110, 111, 117, 120, 133, 166
Syme, K. 147, 172, 178

Tagoilelagi, F. 25
Teale, W. H. 7, 26, 39, 65, 88, 92, 104, 105, 108, 110, 111, 117, 120, 133, 166
Teberosky, A. 51, 126, 133
Thackery, S. 147, 172, 178
Tharp, R. G. 5, 32, 166
Tizard, B. 98, 152, 181, 197
Torrance, N. 142
Troughton, J. 69, 106
Turoa, L. 132

Valdez-Menacha, M. C. 88, 101
Valsiner, J. 2, 14, 15, 17, 30, 35, 56, 77, 78, 79, 116, 123, 201
Van der Veer, R. 14, 56, 77, 116
Vasconcellos, V. M. R. 79
Vygotsky, L. S. 14, 64, 74, 77, 201

Wade, W. R. 96
Wagemaker, H. 35
Wagner, D. A. 94
Walkerdine, V. 122
Wells, G. 19, 71, 104
Wendt, A. 103
Wenger, E. 123
Wertsch, J. V. 14, 36, 74, 77, 78, 127
White, B. L. 104
Whitehurst, G. J. 88, 101
Wiley, C. 159, 160
Wilson, P. T. 159
Wolf, S. A. 56, 90, 122, 141
Wolfgramm, E. 31, 33, 183
Wood, D. 9, 12, 45, 65, 84, 91, 200
Wozniak, R. H. 9

Yensen, H. 35

Index

activities
   ambient 3, 6, 10, 20–5, 28–9, 59, 76, 98–9, 101, 141, 162, 180
   and cognitive endowments 46
   cultural and social voices 30–3
   culturally valued 62, 70
   and descriptions 154–5
   and development of systems 8, 15, 55, 64
   direct family involvement 8, 15
   and expertise 3, 9–10, 14, 15, 38
   and family literacy 3, 5–8
   as a framework for learning 8
   and goals 7, 33, 63, 68, 83, 99, 100, 101, 124–6, 136, 141
   individual 8, 110, 116
   joint 3, 5, 10, 20, 25–9, 59, 62, 64, 74, 76, 141, 162, 168, 170, 173–4
   labelling 28, 46, 86–9, 94, 97, 98, 99, 125
   literacy 23, 38, 51, 62–3, 70–1, 74, 78, 85, 111, 150, 169–73, 180–2, 186–91, 193
   and patterns 46, 103, 111
   personal 3, 6, 10, 14, 15, 20, 29–30, 59, 76, 141
   and rules for participation 7, 33, 63, 74, 99, 141
   and socialisation 14, 105–8, 111
   and settings 11, 161–2
   and tutorial systems 63–4, 69–71, 100, 103, 126–33
   ways of carrying out 33, 99, 141
   writing 5, 7, 11, 21, 24, 25, 29–30, 34, 59, 62, 78, 100, 124–42, 154–5, 172, 191–2
algorithms 45, 60, 128, 136
   letter-production 129–33
alphabet, teaching 26–7, 97, 135
alphabetic routines 128–9
appropriation 60, 77

Bible reading 25, 26, 115
books
   meanings in 9, 88, 89, 105–10, 111
   message 95–6, 108–110
   picture 86–9, 101
   reading 3, 8, 9, 25, 32, 34, 68, 82, 83–6, 97–8, 104–21, 197
   story
      collaborative participation 69–70, 87, 105–10, 111, 114, 117, 119, 176
      and expertise 103, 104, 110–11
      as an interactive system 8, 90–1, 106, 111
      and narrative meaning 89–92, 97, 104–10, 119
      and pedagogical dexterity 112–13
      for performance 92–5, 113–16
      reading 26, 31–3, 85, 86–92, 103–21, 176, 197
   structural properties 97
   see also stories, written; reading

C.A.P. *see* print, concepts
child rearing 17, 18, 53, 110, 122, 169, 171, 193
children
   as active learners 13, 15, 46, 49, 55
   boundaries 18
   construction of knowledge 14
   and conversation 70, 87, 90, 105–6
   development (co-constructivist theory) 15, 51, 53, 54
   differences between 150–2, 154, 159
   and their environment 49
   learning and development systems 8, 48–9, 51, 55, 114, 118, 149–58, 195
   and moment by moment interactions 8, 76, 106
   as performers 70–1
   play and experimentation 15, 30, 48, 55
   and problem solving 8, 13, 44, 45, 46, 49–51, 64, 65, 67, 108, 114
   and qualitative thinking 12
   self-initiated questioning 46, 108
   views of 52–4
   see also education
co-construction
   processes 76, 175
   theory 2, 12, 14–16, 51, 53, 54, 55, 195
cognitive
   development theory 12, 46, 58
   endowment 46, 143
collaborative, participation 69–70, 73, 83, 85, 87, 111, 117, 125
   and authority 108–10, 114–15
   pedagogy and culture 105–10
   and role of the individual 110, 116
   and writing 137–8
collaborative programmes 118, 176
connections
   home-school 184–92
   and settings 164–5, 167–76, 180, 184
constituent elements 42, 48
construction
   and development 14–15
constructivist theory 12, 14, 133
control, gaining 42–4
conversational partners
   children as 70, 87, 90
   and reading 105–6
conversational routine 73
culture
   and family literacy 5, 6, 7, 10, 14, 17, 19, 30–4, 70, 84–5, 104–10, 111, 113–18, 137–8, 142, 149–52, 157, 169–70, 175–6, 182–4, 195, 197
   and models of development 12, 52, 53
   and pedagogy 105–10, 113–16

213

cultural
    capital 18, 170
    forces and Piaget's theory 14
    identity 19, 114–15
    language 84
    meaning 15, 19, 31–3, 59, 85, 111, 117, 123
    processes 7
culturally valued activities 62

descriptions, writing 136–9
development
    and activities 15, 64, 70, 85, 100
    of children (co-constructivist theory) 15, 51, 53, 54, 55
    child's role 38, 39–49, 55, 63, 70, 114
    and co-construction theory 2, 12, 14–16, 51, 53, 54, 55, 195
    and construction 14–15
    concepts 13, 14–15
    defined 2
    and environment 15
    of expertise 9, 10, 38, 39–49, 59–60, 68, 76, 89, 92, 94–5, 96, 133–5, 136, 147, 157, 167
    fixed 15
    human 10
    language 7–8, 62, 64–8, 70, 76, 94, 101
    literacy 7–8, 9, 10, 15, 19, 58–9, 70–1, 94, 104–5, 117, 146, 149, 153, 162, 167, 171–2, 176, 193–5
    models and role of culture 12, 14, 52–4, 162–4
    preschool 7, 9, 31–3, 104, 150, 152–5, 157, 180–2
    reading 7, 11, 82, 85, 98, 100, 101–2
    and relationship between settings 10, 11–12
    socialisation
        model 2, 3, 10, 52–3
        process 8, 10
        and settings 10, 11, 12
    system 3, 4, 8–9, 11, 12, 38, 55, 62–75, 76
    theory 12–16, 52–4
    unitary sequence 15
    verbally controlled 70
    writing 7, 39, 55
dexterity
    and parent education 118–20, 176
    pedagogical 112–13
display routine 73

education
    classroom
        learning 166
    settings 32, 55, 166
    early-childhood 149, 150–1, 154, 172, 179, 180–2, 196
    experiences 30
    and expertise 147, 148–52, 155–7, 160, 185–91
    and home-school connections 184–92, 193, 195–6
    settings 176, 177, 178, 180–5, 195–6
    and socialisation 10
    system and families 21, 164–76, 184–92, 193–6
    and teachers 169–73, 185
    and testing 147–8, 159
education, parent
    and dexterity 118–20, 176
educational resources 34
elements
    control of 42
    constituent 42, 48
embedding skills 42, 65, 83, 89, 92, 97, 101, 128, 133
emergent literacy
    defined 7
    development 7, 12, 59, 55
    and early-childhood education 181
    family system, role 3, 17–34, 59, 76, 116, 119, 195
    and formal instruction 7
    socialisation model 3, 16
emergent writing 171–2
exosystems 164, 193–5
experimentation
    and play 15, 30, 48, 55
    and social experiences 8
expertise
    concept 10
    development 9, 10, 15, 38, 39–49, 59–60, 68, 76, 89, 92, 94–5, 96, 100, 133–5, 136, 147, 157, 167, 185
    on entry to school 148–52, 185
    family-based 10, 62, 10, 104–11
    literacy 15, 38, 172, 185–92
    meaning 44–5
    and personal action 15
    reading 15, 100, 104, 110–11, 114, 118, 154
    and skill 45
    writing 134, 172
expertise, situated
    and activities 9, 10
    defined 9
    in learning and development 3, 4, 9–10
    in written language 10, 111
expositions, writing 136–9

families
    low-income 111, 198
    middle-class 31, 105, 111, 116, 122, 127, 148
    relationship with education system 21, 164–76, 184–92, 193–5
family
    in context 162–4
    cultural meanings 30–3, 59, 84–5, 104–10, 111, 113–16
    education 118–20
    functions of 3, 30–3, 17–19, 111, 118
    support networks 18
family literacy
    activities 3, 5–8, 15, 23, 31–3, 55, 59, 85, 100, 104–5, 111, 154–5, 158, 167–8, 170, 175–6, 180–2, 186–91, 193–5, 197
    direct teaching 119–20
    homework 4, 23, 186–92
    learning and development system 8, 16, 55, 100
    practices 3, 4–5, 17–33, 34, 59, 100, 104, 111, 118–20, 137–8, 142, 149, 151, 157, 182–4, 193–5
    social and cultural identities 5, 6, 7, 10, 14, 17, 19, 30–4, 70, 84–5, 104–10, 111, 113–18, 137–8, 142, 149–52, 157, 169–70, 175–6, 182–4, 195, 197
    and systems 8, 17, 33
    time and resources 5, 18, 31–3, 104–10, 111, 157, 186–92
    and wider environment 10
feedback and performance 93
frames and writing messages 139–40

goals
    and literacy activities 7, 33, 63, 68, 83–4, 90, 93, 99, 100, 101, 105, 136, 141
graphemes 45, 98

homework and family literacy 4, 23, 186–92
human development 10

immersion 67
imitation and performance 93
Initiation-Response-Evaluation sequence 72, 74, 87
interactions
    and cultural meanings 15, 31–3, 70, 84, 85
    focus of 69–70, 106, 111
    language 84
    moment by moment 8, 76, 106

interactions (*cont.*)
  narrative 68–70, 90–1, 106
  patterns 82, 83, 84–5, 90, 106, 111, 125, 128
  and social meanings 14–15, 70, 89, 92–6, 106, 111
internal variation 42
internalisation 60, 77
inventive practice 48, 118–20, 176

knowledge
  children's construction of 14
  of concepts 98, 101–2, 151
  of how stories work 110–12
  and performance 44, 45–6, 114
  and reading 110–12, 114
  and skill 45, 48
  ways of constructing 12, 14, 30
  and writing 45, 58, 102

labelling activity 28, 46, 86–9, 94, 97, 98, 99, 125
language
  concepts 102
  development 7–8, 62, 64–8, 70, 78, 84, 94, 101, 102
learning
  and activities 8, 46, 52–4, 70–1, 72, 84, 100, 167–8
  acts of 46–9, 108
  child's role 38, 44, 46, 49–51, 55, 63, 70, 72–4, 114
  classroom 166
  frameworks 8
  group 32
  personalised 32, 110, 116
  and reading 9, 13, 100, 104–5
  role of cultural and social forces 14, 30–4, 53–4, 70, 84–5, 104–10, 111, 113–16, 197
  and socialisation processes 8
  strategies 46–8, 133, 170–3
  system 3, 4, 8–9, 11, 38, 48–9, 51, 55, 62–75, 76, 195
letter
  forming 135–6
  identification 97, 102, 106, 148, 149, 150–1, 152
  production 130
  sequencing of 131
  sound relationship 135, 140
  writing 5, 24, 25, 129, 136, 140, 149
letter-production algorithms 129–33
literacy
  activities 23, 38, 51, 62–3, 70–1, 74, 78, 85, 111, 150, 169–73, 180–2, 186–91, 193

  and the church 21, 22–3, 25, 26, 92–3, 115, 182–4
  collaborative forms of 23, 69–70, 87, 105–8, 111
  community forms of 23
  concept 7, 176
  development 7–8, 9, 10, 15, 19, 38, 52, 58–9, 70–1, 94–5, 100, 117, 118, 149–50, 153, 154–5, 157, 162, 167, 171–2, 193–5, 197
  defined 7
  formal instruction 7
  leisure 24
  mismatch 84, 101
  occupational 24
  practices 5, 7, 19, 20, 24, 33, 120, 151, 158, 193, 197
  and reading 7, 31, 186–7
  and role of individual 32, 110, 116
  and settings 11, 161, 167, 180–4, 197
  and symbol systems 7
  standards 117–18, 151
  see also emergent literacy; family literacy; stories, written

macrosystems 164
matched settings 166–7
materials
  reading 85–6
  writing 125–6
Matthew Effect 157
meaning
  in books 9, 88, 89
  comprehending 111
  cultural and social 15, 19, 59, 87, 111, 117, 123
  of expertise 44–5
  narrative 86, 89–92, 104–10, 111, 118
  negotiation of 108–10
  network 19
  personal 15
  and role of individuals 110
memory, recitation 94, 95
mesosystems 164, 168, 172, 176, 193
  before school 180–4
messages, making 139–40
microsystems 163, 164

name production 130
names, writing 45–9, 125–35, 141, 142, 157, 160, 180
narrative
  meanings 86, 89–92, 106–10, 111, 118

  reading for 89–92, 104–10, 118–20, 157, 160, 166
  writing 136–9
New Zealand
  literacy in 19, 31–2, 52–4, 95, 104, 105–8, 113, 118, 147, 148–50, 159, 166, 171, 180, 181, 187, 189, 197

parents and literacy activities 169–73, 197
Parents as First Teachers Programme 104
participation
  rules and literacy activities 7, 33, 63, 70, 74, 77, 99, 141
  and self-regulation 114
  see also collaborative participation
patterns
  of discourse 166
  of interactions 82, 90, 106, 111, 125, 127
  learning and development 68, 85
  matching 46
peer groups 75, 173–6
performance
  and knowledge 44, 45–6
  and reading 92–5, 112–16
  routine sequence 73, 93
  tutorials 70–1, 73, 85, 92–4, 113–16
personal
  activities 3, 6, 10, 14, 15, 20, 76, 141
  forces
    learning and development system 8
  identity 19
  meanings 15
  systems 62, 67, 74–5, 89, 92, 94–5, 100, 133–5, 136, 173
phonemes 45, 98, 111
phonemic
  awareness 133, 140
  routines 129
picture-books *see* books
play and experimentation 15, 30, 48, 55
practices
  and family literacy 3, 4–5, 17–33, 34, 59, 100, 104, 111, 118–20, 137–8, 149, 151
  literacy 5, 7, 19, 20, 24, 33, 120, 158
print
  ambient 28–9
  concepts 97–8, 106, 111, 148, 149, 151, 152, 153
problem solving
  by children 8, 13, 44, 45, 46, 49–51, 64, 65, 67, 108, 114

INDEX

reading
  activities 5, 7, 9, 21, 25–6, 29–30, 31, 82, 97, 98, 154, 186–91, 197
    ideas and goals 83–4, 100, 101, 105
      interaction patterns 84–5, 106
    materials 85–6
    tutorial types 85
  collaborative participation 69–70, 73, 83, 85, 105–8, 111, 114
  concepts 97–8
  and decoding 50
  development 7, 11, 68, 82, 85, 98, 100, 101–2, 104–5, 111, 114
  expertise 15, 100, 104, 110–11, 114, 118, 154
  framework 82
  at home 11, 21, 25–6, 34, 55, 104–11, 186–91, 197
  independent 23
  for items 97–8
  for labels 46, 86–89, 97, 98, 99
  and learning 9, 13, 100, 104–5
  and literacy 7, 31, 186–7
  for messages 95–6, 100
  for narrative meanings 89–92, 104–11, 118–20, 166
  personal pleasure 21
  and pedagogical dexterity 112–13
  to preschoolers 31–3, 34, 69, 104–5, 150
  religious purposes 21, 22–3, 25, 26, 92–3, 115, 182–4
  self-regulation 114
  at school 104–5, 111, 117, 121, 135
  signs 13, 28, 46–7, 98–9
  silent 101
  tutorials 82, 85, 86–94, 95–6, 97–8, 116–17
  see also books; stories, written
Reading for Performance 92–5, 112–16, 127, 183, 184
Reading Recovery 101
reasoning stages 12
recitation memory 94, 95
relationships between settings 4, 10–12
rhymes 86, 92–3

scaffolds 65, 66, 68, 70, 74, 111, 117, 139
schools *see* education
settings
  and boundary conditions 11, 12, 18, 166
  classroom 32, 55, 166
  connecting 164–5, 167–73, 195–6
    and peers 173–6, 196
  co-ordinated 12

  educational 176, 177, 178, 180–4
    and literacy 11, 176, 180
  matched 166–7
  and relationships 4, 10–12, 169–73
  socialisation
    and families 10, 14, 32, 161–2, 164, 179
  transfer across 167
  and transitions 164–5
signs
  and literature 13, 28, 46–7, 98–9
skill
  and expertise 45
  and knowledge 45, 48
  performance 44–6
  reading and writing 135, 139
social
  experiences and experimentation 8
  forces and Piaget's theory 14
  identity and family literature 5, 6, 7, 10, 14, 17, 19, 30–4, 70, 84–5, 104–10, 111, 113–18, 137–8, 142, 149–52, 157, 169–70, 175–6, 182–4, 195, 197
  interactions 14–15, 106, 111
  meaning 15, 19
socialisation and families 3, 17, 32, 33, 53–4, 84, 105–8, 111
socialisation model
  components 3
  concept 10, 14
  and development 9, 52–3
  emergent literacy 3, 17
  and family role 5, 17, 18–19
  propositions 5, 7, 8, 9, 10, 12, 19, 33, 55, 76, 99–100, 120–1, 141, 157–8, 176–7
  and settings 10
socialisation patterns 33
socialisation processes
  in development and learning 7, 8, 62
  and family-based expertise 10, 105–8, 111
socialisational settings
  and development 10, 32, 33, 157
socialisations of literacy (SOL) 19, 23–4, 25, 27, 29, 31, 53, 90, 97, 99, 112, 127, 149, 182, 185, 192, 198
spelling 131
  invented 48–9, 140, 143
stories, written
  knowledge of how they work 110–11
  strategies for understanding 3, 111
storybooks *see* books
symbol systems
  and flexibility 48–9
  and generation 49
  and literacy 7, 48–9

systems
  learning and development 3, 4, 8–9, 11, 38, 48–9, 51, 55, 62–75
  personal 62, 67, 74–5, 89, 92, 94–5, 100, 133–5, 136, 173
  symbol 7
  tutorial 62–74, 86–9, 92–4, 100, 126–33, 135–8, 139–40

teachers and literacy activities 169–73, 195–6
text
  as an authority 93, 95, 108–10
  narrative structure 106
transitions and settings 164–5
tutorial systems
  activity centred 63–4, 86–9
  collaborative participation 69–70, 73, 83, 85, 105–8, 111, 125, 138
  defined 62–3
  directed performance 70–1, 73, 85, 92–4, 113–16, 125
  features 63–8
  item conveyancing 71–2, 73, 74, 83, 85, 97, 125
  reading 82, 86–95, 97–8, 100, 112
  relationship-based 68
  and scaffolding 65, 66, 68, 70, 111, 117
  specific configuration 67
  tutorially configured 64–8, 72–4, 113, 116–17
  types 68–72, 83, 85, 125
  writing 126–33, 135–8, 139–40

vocabulary growth 89, 101

word
  games 86
  identification 97
  meaning 89, 101
  segmentation 45, 60
  strategies 60
  written 148, 149, 151
writing
  activities 5, 7, 21, 24, 29–30, 59, 78, 124–42, 154–5, 172, 191–2
  before school 26–7
  and co-construction 58
  and collaborative participation 137–8
  and constituent elements 42
  descriptions 136–9
  development 7, 39–49, 55
  emergent 171–2
  and encoding 50
  expertise 15, 55, 134, 136, 154, 157
  expositions 136–9

writing (*cont.*)
   flexibility 48–9
   at home 25–30, 39–45, 124–41, 142, 143, 157, 160, 191–2
   ideas and goals 124–5, 136
   independent 23
   interaction patterns 125
   and knowledge 45
   letter 5, 24, 25, 129, 136, 140, 149
   materials 125–6
   name 45–9, 125–35, 141, 142, 157, 160, 180
   narratives 136
   parts and wholes 39–44
   processes 55, 58
   at school 135, 151
   single letter 42
   strategies 58
   tutorial 126–33
   see also spelling

written language
   activities 11, 25, 34, 59, 62, 100
   and children 8, 10, 15–16, 19, 25, 110–11, 114
   and co-constructing expertise 16, 58
   concepts 97
   development 15, 19, 55, 58–9
   and expertise 110–11, 157
   and families 17, 18–19, 20, 25, 34, 59, 105, 111
   social and cultural purposes 5, 59, 111
   use of 4–5, 100, 157

Zone of Proximal Development (ZPD) 64, 65, 66, 74, 75
Zones of Freedom of Movement (ZFM) 17, 18, 20
Zones of Promoted Actions (ZPA) 17–18, 97, 105, 172,